Springer Aerospace Technology

The *Springer Aerospace Technology* series is devoted to the technology of aircraft and spacecraft including design, construction, control and the science. The books present the fundamentals and applications in all fields related to aerospace engineering. The topics include aircraft, missiles, space vehicles, aircraft engines, propulsion units and related subjects.

More information about this series at http://www.springer.com/series/8613

Bestugin A.R. · Eshenko A.A. ·
Filin A.D. · Plyasovskikh A.P. ·
Shatrakov A.Y. · Shatrakov Y.G.

Air Traffic Control Automated Systems

 Springer

Bestugin A.R.
Saint Petersburg, Russia

Eshenko A.A.
Moscow, Russia

Filin A.D.
Saint Petersburg, Russia

Plyasovskikh A.P.
Leningrad, Russia

Shatrakov A.Y.
Moscow, Russia

Shatrakov Y.G.
Saint Petersburg, Russia

ISSN 1869-1730   ISSN 1869-1749   (electronic)
Springer Aerospace Technology
ISBN 978-981-13-9388-4   ISBN 978-981-13-9386-0   (eBook)
https://doi.org/10.1007/978-981-13-9386-0

This Springer imprint is published by the registered company Springer Nature Singapore Pte Ltd.
The registered company address is: 152 Beach Road, #21-01/04 Gateway East, Singapore 189721, Singapore

Reviewers

The learning guide discusses the operational principles of ATM automated system, introduces the scientific fields for the simulators for the ATC staff training, and describes the performance of the radiotechnical aids necessary for positioning aircraft by ATM automated systems.

The learning guide is dedicated for studying by senior and post-graduate students of radiotechnical universities in the following specialties: aviation radars, radio navigation, navigation and ATC, data processing and management automated systems, avionic systems, system engineering, data processing and management, and radio engineering.

Anatoly Kozlov, Dr. Sci. in Physics
and Mathematics, Professor, Honored Worker
of Science and Technology of the Russian Federation

Vladimir Aldoshin
Dr. Sci. in engineering, Professor

Sergey Kolganov
Dr. Sci. in engineering, Professor

Introduction

In the frame of the federal target programs, the advanced automated systems (AASs) with aviation control centers are developed and deployed in different regions of the country. For example, in the Moscow region a center is being developed with the area of responsibility covering several hundred aircraft that need to be controlled for takeoff, landing onto four international aerodromes and flights along the airways. At the same time, abroad as well as in our country the fleet of general aviation (GA) is growing, and that is also controlled through an automated system. Currently, in the USA only the GA fleet has already reached 200 thousand aircraft. This learning guide (that will undoubtedly be available for students, post-graduate students, and professionals) at a high professional and scientific level discusses issues such as planning of airspace use, providing of air traffic services, personnel actions in the event of emergency situations, methods that increase the efficiency of automated traffic control systems, algorithms and programs that mitigate conflict situations, methods of air traffic controller simulator training, and functioning and capabilities of automated simulator workstations.

This learning guide also includes the following topics: principles of operation of all radiotechnical systems, their characteristics, and procedure of their use in automated aviation control systems—all to the extent required for students, graduate students, and professionals.

This learning guide will be useful to a wide range of post-graduate students whose scientific research covers the solution to problems and the development of proposals to improve the efficiency of existing radiotechnical systems and radiotechnical systems under the development, as well as structures of general purpose including special fields such as automated systems of information processing and management, navigation and air traffic control, radio-electronic systems, system analysis of data management and processing, radio engineering, radar, and radio navigation.

The creation of domestic ATC AS addresses the task of the government for the development of innovative economy in the country and creation of 25 million new job positions. As practice shows, the commissioning of just one ATC-integrated AAS center allows creating of hundreds of new job positions with a high degree of automation of all production processes. Only highly qualified professionals will be employed for these job positions whose training is targeted by dedicated universities of the country.

   This learning guide includes 11 chapters covering all aspects of the development and operation of ATC AS. The authors of this learning guide are highly qualified professionals who are engaged in the development and implementation of AAS.

Contents

Abbreviations

A/C	Aircraft
A/D	Aerodrome
AAS	Advanced automated system
ABI	Advanced boundary information
AC	Area center
ACAS	Airborne collision avoidance system
ACC	Area control center
A-CDM	Airport Collaborative Decision Making
ACI	Area of common interest
ACRS	Airfield control radar simulator
ACS	Approach control servicing
ACT	Activation message
ADC	Air data computer system
ADF	Automatic direction finder
ADS	Automatic dependent surveillance
ADS-B	Automatic dependent surveillance—broadcast
ADS-C	Automatic dependent surveillance—contract
AeroAI	Aerodrome ATC instructor
AeroFOD	Aerodrome flight operations director
AFDS	Aerodrome flight dispatch supervisor
AFL	Aircraft flight level
AIDC	ATS interfacility data communication
AIP	Aeronautical information publication
AIRMET	Information concerning enroute weather phenomena which may affect the safety of low-level aircraft operations
AMAN	Arrival management
APPR	Approach
AreaAI	Area ATC instructor
ARP	Aerodrome reference point
ARSU	Auxiliary runway supervisory unit

ASD	Air situation display
ASGS	Air situation generator server
ASR	Airport surveillance radar
ASTPM	Automated system of technological process management
ASU	Aerodrome supervisory unit
ATC	Air traffic control
ATD	Actual time of departure
ATFM	Air traffic flow management
ATIS	Automatic terminal information service
ATM	Air traffic management
ATS	Air traffic service
ATS SS	ATS surveillance system
AU	Airspace user
AutoTS	Automated training system
AWS	Automated workstation
AWS-C	AWS-controller
AWS-OP	AWS-operator pilot
AWS-SA/TMM	AWS-system administrator/technical management and monitoring
AWS-TS	AWS-training supervisor
CC of POFL	Counter-Crossing of Flight Level Previously Occupied by another aircraft
CCPM	Control command processing module
CFL	Cleared flight level
CHG	Modification message
CMM	Channel memory module
CompM	Compression module
ConvM	Conversion module
COP	Coordination point
CPDLC	Controller–pilot data link communications
CR	Crossing route
CRA	Conflict resolution advisory
CSU	Circuit supervisory unit
CWP	Controller work procedure
DDL	Digital data link
DDM	Database documentation and replay module
DFZ	Danger flight zone
DL	Data link
DMAN	Departure management
DME	Distance measuring equipment
D-VOLMET	Data link VOLMET
ECM	Electronic computing machine
EPP	Exercise preparation package
ETA	Estimated time of arrival
ETO	Estimated time of overflight

Eurocontrol	European Organisation for the Safety of Air Navigation
FAP OPIVP	Federal Aviation Regulations on the Organization of Airspace Use Planning in the Russian Federation
FAP OrVD	Federal Aviation Regulations on the Air Traffic Management in the Russian Federation
FAS	Final approach segment
FATA	Federal Air Transport Agency of the Russian Federation
FCA	Flow constrained area
FDD	Flight data distribution
FDPS	Flight data processing system
FDU	Flight data updating
FFP	Filed flight plan
FIC	Flight information center
FIS	Flight information service
FP IVP	Federal Rules for the Airspace Use in the Russian Federation
FR	Federal rule of the Russian Federation
FSU	Final supervisory unit
GA	General aviation
GApC	Group Approach Center
GC	General concept
GDB	Graphic database
GLONASS	Global Navigation Satellite System
GNSS	Global navigation satellite system
GPS	Global Positioning System
GS	Glide slope
GS REC	Glide slope receiver
HF	High frequency
HMI	Human–machine interface
IAF	Initial approach fix
ICAO	International Civil Aviation Organization
IFF	Identification friend or foe
IFPP	Initial Flight Plan Processing
IFR	Instrument flight rule
ILS	Instrument landing system
IM	Inner marker
INS	Inertial navigation system
IRD	Integrated radar data
IS	Integrated system
ISC	Integrated system complex
ISS	Integrated system simulator
LAM	Logical acknowledgement message
LAN	Local area network
LCT	Landing control tower
LCU	Line-up control unit
LM	Locator middle

LMK (LMB)	Left mouse key (button)
LMM	Locator middle marker
LoA	Letter of agreement
LOC	Localizer
LOM	Locator outer with radio marker
LRSD	Landing radar simulator display
LSC	Loud speaking communication
LSSU	Landing system supervisory unit
LSU	Local supervisory unit
MAC	Message for Abrogation of Co-ordination
METAR	Aerodrome routine meteorological report (*in meteorological code*)
MKR	Marker receiver
MLS	Microwave landing system
MONA	Monitoring aids
MSA	Minimum safe altitude
MSRC	Means of simulated radio communication
MSSC	Means of simulated service communication
MSVC	Means of simulated voice communication
MTCD	Medium time conflict detection
MTI	Moving target indication
MVT	Movement
NAVSTAR	Navigation satellite providing time and range
NCW	Non-conformance warning
NOTAM	Notice to airman
NOZ	Normal operating zone
NS	Next sector
NTZ	No transgression zone
OAT	Outside air temperature
OFIS	Operational flight information service
OLDI	Online data interchange
OM	Outer marker
OP	Operator pilot
OS	Operating system
P	Planning
PAC	Preliminary activation message
PAR	Precision approach radar
PC	Planning controller
PCU	Procedural control unit
PDM	Control panel display module
PEL	Planned entry level
PIC	Pilot in command
PID	Planning information display
PK	Proficient knowledge
PPD	Potential problem display

PSR	Primary surveillance radar
RA	Resolution advisory
RAD	Radiotechnical Communications and Navigational Equipment Department (*an airport service*)
RATS	Russian Academy of Technical Sciences
RB	Relative bearing of radio station
RBS	Radar beacon system
RCS	Regional Control Servicing
RCU	Radar control unit
RDD	Reference data display
REV	Revision Message
RF	Radar facility (*only within the current document, Chapter 7*)
RF	Russian Federation
RFL	Repetitive flight level
RFP	Repetitive flight plan
RLSU	Remote local supervisory unit
RMK (RMB)	Right mouse key (button)
RNAV	Area navigation
RPIPS	Radar and planning information processing server
RPMS	Remote proficiency maintaining system
RRP	Radio rules and phraseology
RSU	Runway supervisory unit
RTE	Radiotechnical equipment
RVR	Runway visual range
RVSM	Reduced vertical separation minima
RWY	Runway
SAeroSC	Senior aerodrome shift controller
SAreaSC	Senior area shift controller
SCU	Start control unit
SDC of POFL	Same-Direction Crossing of Flight Level Previously Occupied by another aircraft
SDR	Same direction route
SFL	Same flight level
SI	International System of Units
SID	Standard instrument departure
SIGMET	Information concerning enroute weather and other phenomena in the atmosphere that may affect the safety of aircraft operations
SK	Solid knowledge
SLECP	Simulated lighting equipment control panel
SLOP	Strategic lateral offset procedure
SMR	Surface movement radar
SMS	Safety management system
SNET	Software aids for flight safety monitoring
SNS	Satellite navigation system

SPECI	Aerodrome special meteorological report (*in meteorological code*)
SRE	Surveillance radar equipment
SRRNS	Short-range radio navigation system
SSC	Selective subscriber call
SSR	Secondary surveillance radar
SSUC	Senior supervisory unit controller
STAR	Standard instrument arrival
STCA	Short-term conflict alert
SUAI	State University of Aerospace Instrumentation
SVAC	Simulator of the visual airfield condition
SYSCO	System supported co-ordination
TAF	Terminal aerodrome forecast
TC	Tactical controller
TCA	Terminal control area
TDTTC	Civil Aviation Terrestrial Data Transmission and Telegraph Communication network
TFL	Transfer flight level
THR	Threshold
TOBT	Target off-block time
TP	Trajectory prediction
TS	Training supervisor
TSA	Temporary segregated area
TSAT	Target start-up approval time
TSU	Taxiing supervisory unit
TTOT	Target take-off time
UHF	Ultra-high frequency
USM	Universal simulation module
UTC	United coordinated time
VAW	Vertical assistance window
VFR	Visual flight rules
VHF	Very high frequency
VNIIRA	JSC "VNIIRA" (Federal Scientific Production Center) All-Russian Scientific Research Institute of Radio Equipment awarded with the Order of the Red Banner (JSC «VNIIRA»)
VOLMET	Meteorological information for aircraft in flight
VOR	Very-high-frequency omnidirectional radio beacon
VSP	Variable system parameter
WS	Workstation
WTC	Wake turbulence condition
ZhAFEA	Zhukovsky Air Force Engineering Academy

Chapter 1
Management of the Airspace Use Planning in the Russian Federation

1.1 General Considerations

The management of airspace use planning covers the following items:

- collecting and processing by the centers of the Unified Air Traffic Management System (Unified ATM System) of information related to airspace use plans, realization of the said plans, as well as of other information related to issuing of permissions and information messages on the airspace use;
- procedures for strategic, pre-tactical, and tactical (current) planning of the airspace use; coordinating of the airspace use with the aim of its allocation by coordinates, time and altitudes between all airspace users (AUs), as well as provision of the air traffic flow management (ATFM);
- cooperation between different centers of the Unified ATM System, and with the ATS (flight supervision) units of different AUs, and with air defense units subject to the monitoring of compliance with the Federal Regulations' requirements.

Head center of the Unified ATM System coordinates the activities of the area centers of the Unified ATM System.

An area center of the Unified ATM System is responsible for strategic, pre-tactical, and tactical planning and coordinating of the airspace use, as well as coordinating of the ATS (flight supervision) units' activities within the Unified ATM System related to the airspace use management.

A regional center of the Unified ATM System, when authorized to develop plans on the airspace use as per the list of the Unified ATM System areas and regions, is responsible for pre-tactical and tactical planning and coordinating of the airspace use, as well as coordinating of activities related to the airspace use management of the following units:

- regional centers and subcenters of the Unified ATM System within its region of responsibility;
- ATS (flight supervision) units of different AUs when they are authorized to work as the state aviation subsidiary ATC units;

© Springer Nature Singapore Pte Ltd. 2020
Bestugin A.R. et al., *Air Traffic Control Automated Systems*,
Springer Aerospace Technology, https://doi.org/10.1007/978-981-13-9386-0_1

- ATS units that provide flight information services (FISs);
- ATS units that provide services for aerodromes and landing sites within its region of responsibility.

Strategic, pre-tactical, and tactical planning of the airspace use, as well as its coordinating shall be provided based on the following information:

- messages on the plan (schedule) of the airspace use, including information on international and domestic flight plans along the routes with the ATS provision or without the ATS provision and on the use of the flow constrained areas (FCA);
- authorizations (issued by the corresponding federal aviation authorities) for international flights and revocations of the said authorizations;
- prohibitions and restrictions on the airspace use;
- authorizations on the airspace use in prohibited and restricted areas issued by the persons whose interest is supported by setting of the said prohibited or restricted areas, and revocations of the said authorizations;
- messages on the aircraft MVT in the airspace;
- messages on starting and ending of the activity related to the airspace use other than flight operations.

In order to support the airspace use planning, a ground aeronautical telecommunication network, public telephone network, limited access telephone and/or telegraphic network, and Internet network are deployed across the centers of the Unified ATM System, as well as receiving of information messages in print format including fax messages is provided.

The airspace use planning in the centers of the Unified ATM System equipped with the dedicated AS shall be provided through the said AS.

The ATS (flight supervision) units of different AUs, legal entities, and individuals who manage and use the airspace shall provide all information related to their activity on the airspace use to the centers of Unified ATM System in full and timely manner.

Coordinating of the airspace use ensures its flexible use and includes the following:

- providing the safety of the airspace use on change of air, weather, and navigation environment via the implementation of the centers of the Unified ATM System authorization for the airspace reallocation with respect to the state priorities;
- providing the timely introduction and removal of prohibitions and restrictions inside the optimal airspace volumes when the said prohibitions and restrictions are related to the temporary effective or local regulations, and to the short-term restrictions;
- providing the possibility to use the flight restricted areas of airspace when these areas are time-limited.

In course of planning of the airspace use, the ATFM service shall be implemented along the routes with the ATC provision and in terminal areas of civil aerodromes including joint civil and military aerodromes.

For the AFTM service provision, the informative and administrative measures shall be taken in the following cases:

– if the traffic flow volume reaches the limit of the capacity declared by the ATS units that provide area traffic service, approach control service and aerodrome control service, or exceeds the said limit;
– if the airspace capacity decreases as the result of restrictions introduced because of the state priorities realization or weather variations or development of other factors that affect the flight safety.

The informative and administrative measures' implementation is intended for the safe and well-disposed air traffic flow maintenance as well as for the ATS units' capacity control.

Strategic, pre-tactical, and tactical planning of the airspace use, as well as coordinating of the airspace use shall be provided by the aviation personnel of the head, area, and regional centers of the Unified ATM System on a 24-h basis; the UTC shall be used.

The airspace within the Russian Federation (RF) as well as outside of the RF, but where the ATM provision is under the RF responsibility, can be classified as follows:

(a) Class A. IFR flights only are permitted; all flights are provided with air traffic control service and are separated from each other. No speed limitations are applicable. A continuous two-way radio communication is required. All flights are performed subject to the ATC clearance, excluding the cases stipulated in FP IVP, item 114 [1];
(b) Class C. IFR and VFR flights are permitted; all flights are provided with air traffic control service. IFR flights are separated from other IFR flights and from VFR flights. VFR flights are separated from IFR flights and receive traffic information in respect of other VFR flights. No speed limitations are applicable. A continuous two-way radio communication is required. All flights are performed subject to the ATC clearance, excluding the cases stipulated in FP IVP, item 114 [1];
(c) Class G. IFR and VFR flights are permitted. No flight separation is provided. All flights receive flight information if requested. The speed limitation is 450 km/h below 3000 m. A continuous two-way radio communication is required for IFR flights. There is no requirement for a continuous two-way radio communication for VFR flights. The ATC clearance is not required for any flights (see FP IVP [1]).

1.2 Organization of Strategic Planning of Airspace Use

Airspace use strategic planning (strategic planning) shall be carried out by the head and area centers of the Unified ATM System two or more days before the date of the airspace use with the aim of coordinating the issues related to the airspace use management during the following operations:

– international and domestic flights along the ATS routes in accordance with the schedules of regular passengers' and/or cargo transportations provided by the operators that have the necessary licenses (air service schedule);

– flights along the routes without the ATS provision or the airspace use other than flight operations, when the permissions for or restrictions of the airspace use are required.

The Responsibilities of the Head Center of the Unified ATM System
At the stage of strategic planning, the head center of the Unified ATM System is responsible for the following:

– coordinating of the airspace use management in relation to the air service schedules, therefore, gaining and generalizing of the necessary information which can be found in repetitive flight plan (RPL) messages for international and domestic flights along the ATS routes and in their respective modifications;
– receiving and generalizing of the information which can be found in the aircraft's filed flight plan (FPL) messages and other plans for the airspace use other than flight operations, and checking it against the specified airspace structure and the formalization requirements;
– making forecast of the airspace demands for air traffic provision for every forthcoming day for which strategic planning is done, and defining relations between the air traffic demands and the capacity declared by the ATS units;
– developing, coordinating, and implementing of the strategic measures on ATFM in line with the declared air traffic demands and capacity of the ATS units, as well as the prohibitions and restrictions that affect the air traffic flows;
– in conjunction with area centers of the Unified ATM System developing and conducting of the preliminary coordination of the conditions for the airspace use for the state aviation and experimental flights along the routes without ATS provision;
– receiving of permissions on flight operations, permissions on activities other than flight operations, and distributing of the said permissions to the area centers of the Unified ATM System as far as they are concerned, as well as other information related to the airspace use management and flight safety.

The Responsibilities of the Area Center of the Unified ATM System
At the stage of strategic planning, the area center of the Unified ATM System is responsible for the following:

– participating in processes of coordination of the airspace use issues related to the passenger and (or) cargo transportation schedules within its area of responsibility;
– receiving and generalizing of the information which can be found in the aircraft FPL messages and other plans for the airspace use other than flight operations, and checking it against the specified airspace structure and the formalization requirements;
– developing and conducting of the preliminary coordination of conditions for the airspace use for the state aviation and experimental flights along the routes without ATS provision in cooperation with regional centers within its area of responsibility;
– participating in the implementation of strategic ATFM measures in line with the declared air traffic demands and capacity of the ATS (flight supervision) units, as well as the prohibitions and restrictions that may affect the air traffic flows;

– gaining and generalizing of the information that can be found in plans (schedules) of the airspace restricted areas when the period of the restriction validity is provided through NOTAM, as well as information on the use of flight routes in vicinity of aerodromes (hubs), and forwarding of the appropriate reports required for NOTAM issuing to the aeronautical information office;
– receiving from the head center and providing to the regional centers (within its area of responsibility) of permissions on flight operations, on activities other than flight operations, as well as other information related to the airspace use management and flight safety.

For strategic planning of international and domestic flights which are conducted along the ATS routes and in accordance with air service schedules, aircraft operators shall submit RPL or FPL messages to the head center of the Unified ATM System.

RPL Message

RPL messages can be used by the Russian aircraft operators in case when the flight scheduled in the air service schedule for the same day(s) during at least ten successive weeks or every day during at least a ten-day period. Moreover, the intended flight shall be operated in accordance with IFR and along the ATS routes.

The RPL message elements shall have a high degree of stability, which means that the following items must remain unchangeable: validity period of the flight plan, days of operation, aircraft identification, aircraft type and wake turbulence category, RVSM approval, departure aerodrome, off-block time, cruising speed(s), cruising level(s), route to be followed and RF border corridor(s) to cross, and destination aerodrome.

RPL message can be used by foreign operators in cases stipulated in FAR for the *Organization of Airspace Use Planning in the Russian Federation* (FAR OPIVP) [2], item 17.1, provided that the above procedure is set by the bilateral international agreement.

If it is necessary, RPLs can be arranged in appropriate listing form by operator. The information, which contains in the RPL listing form and the procedure of its submitting to the head center of the Unified ATM System, shall correspond with the table of the RF MVT requirements; the said Table shall be appropriately approved.

Modifications to RPL listings related to the new flight as well as excluding or modification of the elements that were already included in previous RPL listings shall be implemented after the message about the modification is provided to the head center of the Unified ATM System in the approved format of RPL listings.

FPL Message

Scheduled air service FPL message shall be sent two or more days before the intended flight (as deemed appropriate by the operator) to the head center of the Unified ATM System in the following cases:

– if the frequency and conditions of flights do not meet the criteria of FAP OPIVP [2], item 17.1;
– if the international flight is conducted by a foreign operator with whose state the RF does not have an international agreement on the RPL procedures;

– if it is known that the international flight will be conducted in the RF airspace with modifications against the RPL data.

Additionally, for strategic planning of the international flights which are conducted along the ATS routes and in accordance with the air service schedules, the head center of the Unified ATM System shall use the information received from FATA at the stage of coordination and issuing of permissions.

Strategic Planning of International Flights
Strategic planning of international flights makes allowance for the following conditions:

– the head center of the Unified ATM System has at its disposal FPL and one of the following documents: RF international agreement, permissions for international transportations issued by the RF Government, Ministry of Foreign Affairs of the RF, Ministry of Industry and Trade of the RF, or General Staff of the Armed Forces of the RF;
– FATA shall issue the permission for international transportations in case for the international flight of a foreign aircraft it is required to use the RF airspace out of the ATS routes or the routes that are closed for international operations, or to use the aerodromes that are closed for international operations for take-offs/landings.

Strategic Planning of Flights Out of the ATS Routes
Strategic planning of flights out of the ATS routes excluding international flights and activities related to the airspace use other than flight operations is intended for:

– institution of early measures for providing safety use of other users' airspace;
– preliminary agreement on ATM issues.

Organization of Ferry Flights
When organizing a ferry or off-aerodrome flight of combat or combat-training state aircraft and experimental aircraft designed for the state aviation use, ferry (off-aerodrome) flight routes and the ATS procedures are subject to preliminary coordination with the head or area center of the Unified ATM System.

For the ferry (off-aerodrome) flights' coordination, the operational agencies of the Unified ATM System shall have information on the entire route where for each stage, the distance, the time, and altitude envelopes (flight levels), and landing and alternate aerodromes shall be specified.

When conducting ferry (off-aerodrome) flights in the airspace of three or more adjacent areas of the Unified ATM System, the data given in item 21.1 of FAP OPIVP and the request for these data coordination shall be submitted to the head center of the Unified ATM System.

When conducting a ferry (off-aerodrome) flight in the airspace of one or two adjacent areas of the Unified ATM System, the data given in item 21.1 of FAP OPIVP and the request for these data coordination shall be submitted to the area center of the Unified ATM System where the departure aerodrome is located.

The dates of the ferry (off-aerodrome) flights' data submitting to the head or area center of the Unified ATM System shall be determined as deemed appropriate by the party who organizes the intended ferry (off-aerodrome) flight; however, the said dates shall remain in frames of strategic period of airspace planning.

The item 21.4 of FAP OPIVP does not cover the cases when a ferry (off-aerodrome) flight route or its portions lay in class G airspace and the expected speeds are over 450 km/h.

In this case, the ferry (off-aerodrome) flight's data shall be submitted to the head or area center of the Unified ATM System at least five days in advance of the flight for timely issuing of NOTAM warning on the prohibitions and restrictions inside the uncontrolled airspace that are related to the intended ferry (off-aerodrome) flights.

With regard to ferry (off-aerodrome) flights conducted in the airspace of three or more adjacent areas of the Unified ATM System, the head center of the Unified ATM System in conjunction with the area centers concerned shall analyze the declared route and assess the possibilities of its use, as well as the procedures of air traffic services (operational control) including the radiotelephony primary and secondary frequencies and the ATC transfer points.

With regard to ferry (off-aerodrome) flights conducted in the airspace of one or two adjacent areas of the Unified ATM System, the area center of the Unified ATM System in conjunction with the regional centers concerned shall analyze the declared route and assess the possibilities of its use, as well as the ATS procedures including the radiotelephony primary and secondary frequencies and the ATC transfer points.

The head center of the Unified ATM System shall send the coordinated data of the ferry (off-aerodrome) flight and procedures of air traffic services (operational control) to the ferry (off-aerodrome) flight administrator within two days after receiving of the request.

The area center of the Unified ATM System shall send the coordinated data on the ferry (off-aerodrome) flight and the ATS procedures to the ferry (off-aerodrome) flight administrator within one day after receiving of the request.

Procedures for Restricted Areas Use

At the stage of strategic planning, the procedures of use of restricted areas shall be coordinated, if the restriction validity is specified in NOTAM or published in aeronautical information publications.

In case it is necessary to use a temporary segregated area (TSA) as a gunnery range for target practices, missile firing, bombing, airdrop missions, antihail rockets' firing, working with weapon at their storage facilities, conducting research studies inside the atmosphere, blasting operations, flights inside special zones outside of the aerodromes (helidromes) areas, and unmanned aircraft operations, AUs shall send the TSA use schedule (plan) to the area center of the Unified ATM System at least five days in advance of the intended use.

In the TSA use schedule (plan), the TSA number, the date and time of its airspace use shall be specified.

The TSA use schedule (plan) can be sent with indication of one or more dates of the TSA use, but not more than one month in advance.

In case of use for operations outside of terminal areas of aerodromes in the TSA use schedule (plan), the data related to the air routes to be used shall be additionally specified.

After the TSA use schedule (plan) is received, the area center of the Unified ATM System shall send a request for NOTAM publishing to the aeronautical information office, publish the information on the TSA activation together with the related restrictions set on the ATS routes on the Internet, and inform the head and area centers of the Unified ATM System concerned.

During the pre-tactical period of airspace use planning, it is possible to make some amendments to the TSA activation time in line with the airspace use plan inside of the earlier declared dates.

Coordination of Unmanned Free Balloons' Launch

At the stage of strategic planning, launches of unmanned free balloons (rocket-probes, rawinsonde balloons, pilot balloons, and other material objects) shall be coordinated.

Coordination of the unmanned free balloons' launch if they are launched during the internationally agreed dates with the aim to obtain the weather data on the atmospheric conditions shall be done on base of schedules (extracts of the annual plans).

The Federal Service for Hydrometeorology and Environmental Monitoring regional offices shall send the schedules (extracts of the annual plans) to the area centers of the Unified ATM System and to the Air Force and Air Defense headquarters every year before December 15.

Information on any changes in the launch schedule shall be passed at least 15 days before the estimated date of launch.

Coordination of single launches shall be done based on the airspace use plan which shall be submitted to the area center of the Unified ATM System concerned at least 3 days before the estimated date of launch.

In the meantime, the area centers of the Unified ATM System shall:

– provide launch schedules (extracts of the annual plans) and single launch plans to the appropriate regional centers of the Unified ATM System;
– determine the prohibitions and restrictions for the ATS routes, flight routes in vicinity of the aerodromes for the ATFM provision.

Forecast of the Airspace Use Demands

Based on the received FPL messages and messages of the airspace use other than flight operations, information on the set airspace prohibitions and restrictions, and statistics on the previous air traffic intensity, the head center of the Unified ATM System shall develop the forecast of the airspace demands for the air traffic provision for every forthcoming day.

For the said demand forecast development, the airport technical capacity shall be considered.

Based on the airspace demand forecast for the air traffic provision for every forthcoming day, the head center of the Unified ATM System shall define the relationship between the air traffic demands and the capacity declared by the ATS (flight supervision) units.

In case the demands exceed the available capacity of the ATS (flight supervision) units, the head center of the Unified ATM System in cooperation with the area centers shall develop, coordinate, and implement strategic measures for the ATFM provision, such as:

- informing the ATS (flight supervision) units to let them develop the means to increase their capacity;
- preparing of new recommendations on the airspace management;
- correcting of the spatial and time characteristics of the route declared by an AU and intended to be used in the RF in accordance with the RF state priorities and with consideration of the capacity declared by the ATS (flight supervision) units.

1.3 Organization of Pre-tactical Planning of Airspace Use

Pre-tactical planning of the airspace use is conducted by the head, area, and regional centers of the Unified ATM System one day before the date of the intended airspace use with the aim of allocating the airspace use by the position, time, and altitude (airspace allocation), including allocation of the ATS routes' airspace between the aircraft within the air traffic flow.

The airspace allocation is a process in course of which the centers of the Unified ATM System develop conditions of the airspace use for the forthcoming day based on the received FPL messages related to the flights in classes A and C airspaces, unmanned aircraft operations, plans on the airspace use other than flight operations, as well as the prohibitions and restrictions aimed for the safety of the airspace use.

Based on the pre-tactical planning, the daily plan of the airspace use shall be developed and distributed to ATS (flight supervision) units and the AUs whose activity is not connected with flight operations.

The daily plan contains permissions for and conditions of the airspace use based on which the ATS (flight supervision) unit shall provide the air traffic control clearance (the clearance) upon the PIC's request.

For conducting any activity other than flight operations, permissions and conditions for the airspace use issued by the center of the Unified ATM System concerned act as the basic documents.

Daily Plans of the Airspace Use

The daily plan issued by the head center of the Unified ATM System is a block of information related to the aircraft flight plans, including international and domestic scheduled flights, and permissions for their airspace use for the forthcoming day.

The information on flight plans of international and domestic scheduled flights where strategic planning has been carried out on base of the RPL messages shall be included into the daily plan issued by the head center of the Unified ATM System and distributed to the area centers concerned in the FPL format.

The distribution of the said FPLs shall be completed by the head center of the Unified ATM System not later than 20 h before the estimated aircraft departure time.

The daily plan issued by the area center of the Unified ATM System is a block of information comprising the following items:

- aircraft and unmanned aircraft flight plans;
- plans on the airspace use other than flight operations if conducted within the Unified ATM System coverage (only for the area centers of the Unified ATM System working where no regional centers are established);
- permissions for the airspace use for the forthcoming day in frame of operations and activities stated in the subitem above;
- flight plans and associated permissions for the airspace use issued by the head center of the Unified ATM System as far as the Unified ATM System airspace is concerned.

The daily plan issued by the regional center of the Unified ATM System is a block of information comprising the following items:

- aircraft and unmanned aircraft flight plans, as well as plans on the airspace use other than flight operations, and the associated permissions for the airspace use in accordance with the RF Federal Regulations;
- aircraft and unmanned aircraft flight plans, and the associated permissions for the airspace use issued by the area center of the Unified ATM System as far as the Unified ATM System airspace is concerned.

Tasks of the Head Center of the Unified ATM System
At the stage of tactical planning, the head center of the Unified ATM System is responsible for the following:

- receiving of plans for the airspace use for the forthcoming day and checking them against the specified airspace structure and the formalization requirements;
- developing, analyzing, and coordinating of the airspace use plan for the forthcoming day (daily plan);
- issuing of permissions on the airspace use in accordance with the RF Federal Regulations;
- informing the area centers of the Unified ATM System and the responsible air defense units about the daily plan, the prohibitions for and restrictions of the airspace use issued for the forthcoming day;
- developing, coordinating, and implementing of the pre-tactical ATFM measures in line with the declared air traffic demands and capacity of the ATS (flight supervision) units, as well as the prohibitions and restrictions that affect the air traffic flows.

The Tasks of the Area Center of the Unified ATM System
At the stage of tactical planning, the area center of the Unified ATM System is responsible for the following:

- receiving of plans for the airspace use for the forthcoming day and checking them against the specified airspace structure and the formalization requirements;
- publishing of information about all prohibitions and restrictions issued for the forthcoming day in relation to the airspace classified as "G" use on the Internet;
- developing, analyzing, and coordinating of the daily plan incorporating into it the airspace use plans issued by the head center the Unified ATM System and the adjoining area centers;
- distributing the extracts of the daily plan to the regional centers the Unified ATM System as far as they are concerned;
- issuing of permissions on the airspace use in accordance with the RF Federal Regulations;
- informing the responsible air defense unit about the daily plan of the airspace use issued for the forthcoming day;
- developing, coordinating, and implementing of the pre-tactical ATFM measures in line with the declared air traffic demands and capacity of the ATS (flight super-vision) units, as well as the prohibitions and restrictions that affect the air traffic flows.

The Tasks of the Regional Center of the Unified ATM System

At the stage of tactical planning, the regional center of the Unified ATM System is responsible for the following:

- receiving from the area center of the Unified ATM System of the daily plan infor-mation related to the use of the appropriate area of the airspace within the Unified ATM System;
- developing of the conditions of use of the appropriate area of the airspace where aircraft are operated out of ATS routes, and unmanned aircraft are operated in classes A, C, and G airspace, as well as for the activities other than flight operations;
- issuing of permissions on the airspace use in accordance with the RF Federal Regulations and upon the request;
- making provisions for implementation of the pre-tactical ATFM measures devel-oped by the head and area centers of the Unified ATM System as far as the capacity of the appropriate ATS (flight supervision) units is concerned;
- monitoring of the capacity of the ATS (flight supervision) units concerned, cal-culating the rate changes of the capacity, informing the area center of the Unified ATM System concerned about the said rate changes.

Pre-tactical ATFM Activity

In course of pre-tactical planning, the operational units of the Unified ATM System shall provide the following pre-tactical ATFM for the forthcoming day along the ATS routes:

- defining of the air traffic flow which considers RPL data regarding the forthcom-ing day, as well as the FPL messages submitted for the forthcoming day to the operational units of the Unified ATM System in course of strategic and pre-tactical planning;

- based on the air traffic flow spatial and time parameters checking of the air traffic flow against the capacity declared by the ATS units; by that, the given prohibitions and restrictions must be considered when they may affect the air traffic flows and civil aerodromes' areas.

Allocating the airspace along the ATS routes in course of the air traffic flow planning for the forthcoming day shall be done with respect to the state priorities.

Each FPL message related to the international flight or domestic flight operated within more than one area of the Unified ATM System and submitted to the head center of the Unified ATM System, in case the said flights are intended to be operated along the ATS routes, shall be analyzed against its influence on the ATS units' workload.

The influence of domestic flights on the ATS units' workload, when the flight is intended to be operated within one area of the Unified ATM System, shall be analyzed by the area center of the Unified ATM System.

In case it becomes clear that the air traffic demands exceed the capacity declared by the ATS (flight supervision) units, the head or area center of the Unified ATM System shall develop the pre-tactical measures and implement/customize the said measures (based on agreement with the concerned ATS (flight supervision) units) by the following way:

- by informing of the ATS (flight supervision) unit about the necessity of developing the measures aimed to increase the declared capacity for the forthcoming day;
- by defining in conjunction with the AU concerned of possible changing of the declared flight conditions including the modifying of the route or its time parameters (time of departure or landing).

In course of implementing of the pre-tactical ATFM procedures, the centers of the Unified ATM System and ATS (flight supervision) units shall use formalized messages related to ATS and specified in the table of the RF MVT.

If the pre-tactical ATFM measures are implemented in relation to the intended flight, AU or his designated representative shall take the following steps through the ATS (flight supervision) unit of departure aerodrome:

- cancel the aircraft FPL which was submitted to the head or area center of the Unified ATM System;
- submit a new FPL to the head or area center of the Unified ATM System specifying new aircraft flight conditions which were coordinated with the concerned center of the Unified ATM System.

Regulative measures in course of pre-tactical ATFM planning shall not be applied to the following types of flights:

- flights with "A," "K," and "PK" status;
- search-and-rescue or rescue operations;
- natural and man-made emergency flights, flights for medical, and other humanitarian purposes;
- flights under special conditions;

- abnormal operations;
- flights that were subjected to unlawful interventions;
- state aviation and experimental flights performed from the aerodromes where the procedures and rules defined for civil aerodromes are not applicable.

Pre-tactical Planning of Out of the ATM Routes

In relation to the flights intended to be conducted out of the ATS routes, pre-tactical planning shall be implemented in the following cases:

- for state aviation and experimental flights when the use of restricted airspace areas is intended;
- for ferry and off-aerodrome flights of combat or combat-training state aircraft and experimental aircraft designed for the state aviation use;
- aerial works in the frontier areas;
- aerial works and parachute jumps, as well as demonstration missions above the inhabited areas.

In all cases stated in FAP OPIVP, item 40, AUs shall submit the FPL messages in such a manner that the appropriate authority receives them one day before the intended flight, and they are received:

- by the head center of the Unified ATM System before 14:00 (Moscow time) if the flight is intended to be operated within more than two areas of the Unified ATM System airspace;
- by the area center of the Unified ATM System before 16:00 (local time) if the flight is intended to be operated within one or two adjoining areas of the Unified ATM System airspace.

In case the area center of the Unified ATM System receives the FPL for the state aviation and experimental flights when the use of restricted airspace areas and off-aerodrome area routes is intended, the said FPL data shall be checked against the earlier published appropriate NOTAM.

If it is necessary to correct the data related to the restricted areas and off-aerodrome area routes, AU shall coordinate the said data with the area center of the Unified ATM System responsible for submitting the request for NOTAM publishing to the aeronautical information office.

Issuing of NOTAM related to the restricted area activation time and to the time of the off-aerodrome routes' use can be done within the dates of intended flights.

FPL for the aerodrome traffic shall be included into the daily plan issued by the area center of the Unified ATM System and distributed to the regional center of the Unified ATM System which is responsible for the aerodrome concerned.

Moreover, FPL for the aerodrome traffic shall be distributed to another regional center of the Unified ATM System if the restricted area or off-aerodrome route enters the airspace area which is under the responsibility of that regional center.

Permissions for the airspace use for aerodrome traffic, if it is intended to use the restricted areas or off-aerodrome routes, shall be issued by the regional center of the Unified ATM System and may be distributed to the aerodrome ATS (flight supervision) unit the day before the intended flight.

FPLs for combat or combat-training state aircraft and experimental aircraft designed for the state aviation use, intended for ferry flights out of the ATS routes, at the stage of pre-tactical planning shall be submitted to the head or area center of the Unified ATM System the day before the intended ferry flight.

Data related to the ferry flight routes and ATS procedures to be used specified in the related flight plans shall meet the data agreed at the stage of strategic planning.

In case when the aircraft FPL allows for the subsequent changes of the flight conditions in relation to the earlier agreed air route and ATS procedures, the head or area center of the Unified ATM System shall define and correct the conditions of the airspace use if required.

Permissions for and Conditions of the Airspace Use

Permissions for and conditions of the airspace use for ferry flights of combat or combat-training state aircraft and experimental aircraft designed for the state aviation use shall be distributed by the area centers of the Unified ATM System to the following authorities:

– regional centers of the Unified ATM System and ATS (flight supervision) units who will directly control the ferry flight;
– ATS (flight supervision) units of the departure, destination, and alternate aerodromes along the route.

Permissions for and conditions of the airspace use for aerial works in the frontier areas shall be issued by the appropriate centers of the Unified ATM System at the stage of strategic planning based on FPL and on the permissions issued by the regional body of the Federal Security Service of the RF (FSB).

AU shall submit the copy of the FSB regional body permission to the area center of the Unified ATM System.

Permissions for and conditions of the airspace use for aerial works and parachute jumps as well as demonstration flights above the inhabited areas shall be issued by the operational unit of the Unified ATM System based on the provided FPL which has to be submitted at the stage of pre-tactical planning and based on the permission issued by the appropriate local authority.

AU shall submit the copy of the local authority permission to the area center of the Unified ATM System.

Permissions for and conditions of the airspace use for unmanned aircraft operations shall be developed at the stage of pre-tactical planning based on the FPL message submitted to the area center of the Unified ATM System the day before the intended use excluding the cases when the unmanned aircraft operations are intended to be conducted in classes A or C airspace in accordance with item 114 of the RF Federal Regulations.

By that, FPL for unmanned aircraft operations shall be included into the daily plan issued by the area or regional center of the Unified ATM System at the stage of pre-tactical planning in case if:

– temporary or local procedures are implemented for the said flight;
– the flight is intended to be operated within class A or C airspace where TSAs will be set, and it that would be a defense, governmental, or public security flight, as well as a search-and-rescue flight or relief flight conducted during the natural catastrophes or emergency situations.

Pre-tactical planning in relation to target practices, missile firing, bombing, airdrop missions, antihail rockets' firing, working with weapon at their storage facilities, conducting research studies inside the atmosphere, blasting operations, and activities other that flight operations shall be done based on the plans for the said activities or in accordance with the schedules (annual plans), and only when restricted flight areas or danger areas over high seas are set beforehand.

Pre-tactical planning in relation to airship or lighter-than-air aircraft operations within classes A or C airspace shall be done based on the provided FPL.

At the stage of pre-tactical planning, the operational units of the Unified ATM System shall define information related to antihail rockets' firing, restrictions to be set on the ATS routes, procedures of communication with the operational team which conducts the said firing.

1.4   Organization of Tactical (Current) Planning of the Airspace Use

Tactical (current) planning of the airspace use (tactical planning) is conducted by the head, area, and regional centers of the Unified ATM System in course of the daily plan implementation by means of airspace allocation over position, time, and altitude (airspace allocation) with the aim of ensuring the safety of the planned activity and the activity, the plans for which are submitted during the current day.

The Functions of the Head Center of the Unified ATM System

At the stage of pre-tactical planning, the head center of the Unified ATM System shall cover the following functions:

– receiving of plans for the airspace use for the current day and checking them against the specified airspace structure and the formalization requirements;
– implementing the plan for the current day (current daily plan) and coordinating it with the current air, weather, navigation situation and with consideration of state priorities;
– issuing of permissions on the airspace use in accordance with the RF Federal Regulations;
– informing the responsible air defense unit about the changes and additions to the current daily plan, the prohibitions and restrictions of the airspace use issued for the current day, and coordinating with the responsible air defense unit of the issues related to crossing of the RF state border and to monitoring of compliance with the Federal Regulations;

– implementing and coordinating of the tactical ATFM measures in line with the declared air traffic demands and capacity of the ATS (flight supervision) units, as well as the prohibitions and restrictions that affect the air traffic flows, and the current air, weather, navigation situation.

The Functions of the Area Center of the Unified ATM System
At the stage of pre-tactical planning, the area center of the Unified ATM System shall cover the following functions:

– implementing the plan for the current day (current daily plan) and coordinating it with the current air, weather, navigation situation and with consideration of state priorities;
– receiving of plans for the airspace use for the current day and checking them against the specified airspace structure and the formalization requirements;
– if necessary, updating the information related to the prohibitions and restrictions available for the class G airspace concerned published on the Internet;
– issuing of permissions on the airspace use in accordance with the RF Federal Regulations;
– receiving and distributing to the ATS (flight supervision) units, which provide FIS in the class G airspace, and to the responsible air defense units the appropriate messages on the airspace use;
– implementing of the tactical ATFM measures;
– informing the responsible air defense unit about the changes and additions to the current daily plan, the prohibitions and restrictions of the airspace use issued for the current day, and coordinating with the responsible air defense unit on the issues related to monitoring of compliance with the Federal Regulations;
– developing, coordinating, and implementing of the tactical ATFM measures in line with the declared air traffic demands and capacity of the ATS (flight supervision) units, as well as the prohibitions and restrictions that affect the air traffic flows, and the current air, weather, navigation situation.

If no regional center is established within the appropriate area of the Unified ATM System, the area center of the Unified ATM System in addition shall cover the main tasks on the management of airspace use planning prescribed for the regional center of the Unified ATM System.

The Functions of the Regional Center of the Unified ATM System
At the stage of tactical planning, the regional center of the Unified ATM System shall cover the following functions:

– receiving from the area center of the Unified ATM System of the extract from the daily plan and the current daily plan changes (within its region of responsibility);
– developing of the conditions of use of the appropriate area of the airspace where aircraft are operated out of ATS routes, and unmanned aircraft are operated in classes A, C, and G airspace, as well as for the activities other than flight operations;
– issuing of permissions on the airspace use in accordance with the RF Federal Regulations;

- implementing of the tactical and pre-tactical ATFM measures;
- monitoring the ATS (flight supervision) units' activities when they are implementing the tactical ATFM measures and reporting the results to the area center of the Unified ATM System concerned;
- making provisions for implementations of the pre-tactical ATFM measures developed by the head and area centers of the Unified ATM System as far as the capacity of the appropriate ATS (flight supervision) units is concerned.

Tactical ATFM Activity

At the stage of pre-tactical planning, the provisions for tactical ATFM planning within the ATS routes are based on the daily plan data and on the FPL messages received by the head or area center of the Unified ATM System in relation to the following operations:

- scheduled international and domestic flights;
- international and domestic flights with deviations from the scheduled flights;
- international flights along the ATS routes;
- unscheduled domestic flights along the ATS routes.

FPL Message

Scheduled international and domestic flights' FPL messages shall be received by the head center of the Unified ATM System at least one hour in advance of the intended departure in the following cases:

- if RPL procedure was not applied;
- if the FPL message was not submitted before the tactical planning stage.

The FPL messages about international and domestic flights with deviations from the scheduled flights and the FPL message about a single international flight along the ATS route shall be received by the head center of the Unified ATM System at least three hours in advance of the intended departure.

The following events shall be taken as deviations from the scheduled flights:

- if the flight has departed more than 30 min later against the time specified in the air service schedule;
- if the aircraft identification, aircraft type, and wake turbulence category, RVSM approval, departure aerodrome, off-block time, route to be followed and RF border corridor(s) to be crossed or if the destination aerodrome was changed.

The FPL message for the unscheduled domestic flights along the ATS routes if intended to be operated within more than one area of the Unified ATM System shall be received by the head center of the Unified ATM System at least three hours in advance of the intended departure.

The FPL message for the unscheduled domestic flights along the ATS routes, if intended to be operated within one area of the Unified ATM System, shall be received by the head center of the Unified ATM System at least one hour in advance of the intended departure.

Tactical ATFM Activity

Tactical ATFM activity is provided based on the analysis of the air traffic flows along the ATS routes in accordance with the daily plan and its changes, capacity declared by the ATS (flight supervision) units with consideration to the related prohibitions and restrictions, as well as the current air, weather, navigation situation.

When the head and area centers of the Unified ATM System receive the FPL message, they shall analyze how the said flight affects the ATS (flight supervision) units' workload and identify the units, whose declared capacity can be exceeded.

If it is expected that the declared capacity will be exceeded, the head center of the Unified ATM System shall develop and implement the following ATFM tactical measures:

– informing of the ATS (flight supervision) units about the expected exceedance of the airspace demand against the capacity declared by the ATS units and developing of the appropriate measures aimed to increase the said capacity in cooperation with the ATS (flight supervision) unit;
– providing AU with suggestions to modify the flight conditions declared in the aircraft FPL (such as position, time, altitude).

The information of the tactical ATFM measures shall be distributed to the operational units of the Unified ATM System in the areas of whose responsibility the said measures shall be implemented and to the aerodrome ATS (flight supervision) units in a form of a formalized message in accordance with the approved table of the RF MVT.

If the ATFM tactical measures are implemented subject to the planned flight, the AU shall take the following steps:

– cancel the aircraft FPL, for which a related message was submitted to the head or area center of the Unified ATM System;
– submit a new FPL message specifying the airspace use conditions which were coordinated with a center of the Unified ATM System;
– ensure that the flight is conducted in accordance with the airspace use conditions (such as position, time, altitude) issued by the head or area center of the Unified ATM System.

Permission for the Airspace Use

Permission for the airspace use for the flight operations along the ATS routes shall be distributed by the head center of the Unified ATM System to the ATS units providing the aerodrome control service, approach control service, serving the departure, destination and alternate aerodromes, and providing ATS along the route.

Permission for the airspace use shall be transmitted to the ATS (flight supervision) units serving the aerodrome in a form of a formalized message by the head or area center of the Unified ATM System through the ground aeronautical telecommunication network in accordance with the approved table of the RF MVT.

If the ground aeronautical telecommunication network is not available at the aerodrome, permissions for the airspace use shall be transmitted to the aerodrome

ATS (flight supervision) units by the appropriate head center of the Unified ATM System through the public telephone network.

The rules stipulated in the current subitem shall be used for provision of the state aviation operations along the ATS routes considering that the communication with the units serving the aerodromes used by the state aviation related to departures and landings shall be done in accordance with the regulations established for the state aviation.

Tactical Planning of the Operations Out of the ATM Routes

Tactical planning of the operations out of the ATS routes as well as of the activities other than flight operations shall include:

- specifying, amending, and correcting of the current daily plan in order to ensure provision of the planned activity and of the activity, the plans for which are submitted during the current day;
- issuing of permissions for the airspace use;
- monitoring of the airspace use plans' realization;
- receiving of notifications on the flight and transmitting the said notifications to the ATS units providing FIS and alerting service in class G airspace.

The head center of the Unified ATM System shall supplement its daily plan when it receives:

- at least 6 h in advance of the departure—the FPL message for the flight out of the ATS route if it is an IFR flight (excluding ferry flights out of the ATS routes, as well as following off-aerodrome operations: flights of combat or combat-training state aircraft and experimental aircraft designed for the state aviation use) within three or more airspace areas of the Unified ATM System;
- at least 3 h in advance of the departure—the FPL message for the flight out of the ATS route if it is the VFR flight within three or more airspace areas of the Unified ATM System.

The area center of the Unified ATM System shall supplement its daily plan when it receives:

- at least 3 h in advance of the departure—the FPL message for the flight out of the ATS route if it is an IFR flight (excluding ferry flights out of the ATS routes, as well as following off-aerodrome operations: flights of combat or combat-training state aircraft and experimental aircraft designed for the state aviation use) within one or two airspace areas of the Unified ATM System;
- at least 1 h in advance of the departure—the FPL message for the flight out of the ATS route if it is a VFR flight within one or two airspace areas of the Unified ATM System.

The regional center of the Unified ATM System shall supplement its daily plan when it receives:

- at least 3 h in advance of the departure—the FPL message for the flight out of the ATS route if it is an IFR flight (excluding ferry flights out of the ATS routes, as well as following off-aerodrome operations: flights of combat or combat-training state aircraft and experimental aircraft designed for the state aviation use) within one or two airspace areas of the Unified ATM System;

- at least 1 h in advance of the departure—the FPL message for the flight out of the ATS route if it is a VFR flight within the Unified ATM System airspace or two adjoining airspace regions of one area of the Unified ATM System;
- at least 1 h before the beginning of the aerodrome flights—the FPL message for the said flights subject to the declared operating procedures of the concerned aerodrome (landing site);
- at least 3 h in advance of the unmanned aircraft operations including: defense, governmental or public security flights, as well as search-and-rescue operations, and relief flights conducted during natural catastrophes or emergency situations.

The aircraft FPLs defined in this subitem shall be passed by the area center to the regional centers of the Unified ATM System.

Permission for the airspace use for the flight operations out of the ATS routes shall be distributed to the ATS (flight supervision) units providing the aerodrome control service, approach control service, serving the departure, destination and alternate aerodromes, and providing ATS along the route.

Permission for the airspace use shall be distributed by the head center of the Unified ATM System to the ATS (flight supervision) unit serving the aerodrome in form of a message transmitted through the ground aeronautical telecommunication network in accordance with the approved table of the RF MVT.

If the ground aeronautical telecommunication network is not available at the aerodrome or landing site, permissions for the airspace use shall be transmitted to the ATS (flight supervision) units serving the aerodrome by the head center of the Unified ATM System concerned through the public telephone network.

If the aircraft departure is intended from the aerodrome or landing site where no control serving is provided, the regional center of the Unified ATM System shall issue the permission for the airspace use to PIC upon his/her request before the departure.

The ATS unit serving the aerodrome shall inform the center of the Unified ATM System at least:

- 5 min before the commencement of the planned activity: about its actual commencement, delay, shifting, or cancelation;
- 10 min after the completion of the activity: about its completion, about intervals in aerodrome flights of more than 1 h;
- immediately after expiration of the estimated aircraft landing time or in case the aircraft did not arrive to the destination airport and no information related to its location is available.

AUs shall commence their activity within 30 min from the planned time for the commencement or the time specified by the head center of the Unified ATM System.

If AU failed to initiate its activity during the above period, it shall inform the operational unit of the Unified ATM System and receive corrected conditions of the airspace use or cancel the declared plan on the airspace use.

The ATS unit serving the aerodrome for the state or experimental aviation, if the aerodrome will be used for the intermediate stop, at least 20 min after receiving of the request, shall inform the regional center of the Unified ATM System about the decision of the chief aerodrome flight operations director taken on the aircraft intermediate stop.

Tactical Planning of the Airspace Use Other than the Flight Operations
Tactical planning of the airspace use other than the flight operations shall be conducted based on the daily plan data.

Pilot balloons launch unit shall request permission from the regional center of the Unified ATM System not earlier than 4 h in advance of the planned launch but not later than 2 h in advance.

In case the request for the airspace use or the notification of the launch time shifting was not submitted to the regional center of the Unified ATM System in the specified time, the launch operation shall be prohibited.

The regional center of the Unified ATM System issued to the launch unit the permission for launch at least 1 h in advance of the balloon single launch.

In case the pilot balloon launch is delayed for more than 5 min, the launch may be done only after the additional permission is received from the regional center of the Unified ATM System.

After the launch is conducted, the launch unit shall report to the regional center of the Unified ATM System the actual commencement time, and in addition upon the request the following data: its flight altitude, azimuth, and distance from the launch unit location, as well as the estimated [flight path] deviation.

All operations related to the firing, missile launching, bombing, blasting operations, and other works associated with electromagnetic and other type of emissions, as well as other similar activity shall be conducted in accordance with the daily plan for the restricted flight areas use in order to ensure the safety of the airspace use.

For this type of activity, the regional center of the Unified ATM System shall issue the permission for the airspace use at least 1 h in advance of the commencement of firing, missile, or blasting operations.

Based on forecast and actual hail process developing, the plan for the antihail firing, missile, and blasting operations shall be submitted to the regional center of the Unified ATM System at least 2 h in advance of the estimated time of the commencement of the antihail firing.

When the plan for the firing, missile, and blasting operations is received, the regional center of the Unified ATM System shall develop (or update the existing) short-term restrictions and distribute them to the concerned authorities at least 1 h in advance of the antihail operations.

In case the antihail operation is canceled or the time of its commencement (completion) is shifted the officer-in-charge shall inform the regional center of the Unified ATM System immediately and report the new time of its commencement and completion.

In case of the immediate need for the antihail operations, the officer-in-charge shall request permission for the operations without submitting the firing, missile, and blasting operations' plan.

In this case, the regional center of the Unified ATM System (with consideration to the current air situation) shall immediately set the short-term restrictions and after the withdrawal of all aircraft from the region of antihail operations issue the permission for the antihail operations.

The antihail operations are prohibited in the following cases:

- flight operations and other activities with the purposes of the military and operational tasks aimed to protection of national interests;
- flight operations and other activities with the purposes of spacecraft crew members' searching and evacuation, search-and-rescue operations, and relief flights being conducted during natural catastrophes, accidents, or other emergency situations.

When the decision is taken to conduct the lighter-than-air aircraft operations, the responsible authority shall request from the regional center of the Unified ATM System the permission for the airspace use not earlier than 4 h in advance of the planed operation but not later than 2 h in advance of the launching of the lighter-than-air aircraft along the routes based on the amended weather forecast.

The regional center of the Unified ATM System issues permission for the airspace use not later than 1 h in advance of the planned or corrected flight commencement time.

In adverse weather conditions, the responsible authority shall inform the regional center of the Unified ATM System that its launch (lifting) time is changed at least 30 min in advance of its planned launch (lifting) time.

1.5 Cooperation in Course of the Airspace Use Planning and Coordinating

Cooperation between the head, area, and regional centers of the Unified ATM System shall be maintained in accordance with the standard operating procedures for the executive officers and duty shifts.

Cooperation of the head, area, and regional centers of the Unified ATM System with the ATS (flight supervision) units serving the aerodromes shall be provided by the special structural divisions including the air traffic planning subdivisions which are responsible for the following tasks:

- submitting the FPLs to the operational units of the Unified ATM System, as well as other information related to the aircraft MVT in the aerodrome terminal areas in accordance with the approved table of the RF MVT;
- developing of the daily plan of the air traffic in aerodrome terminal areas, its coordinating with and approving by the centers of the Unified ATM System, including the ATFM issues;

- providing the information on the permissions issued for the airspace use, prohibitions and restrictions of the use of the aerodrome terminal areas and conjoined areas and regions where the ATS service is provided to the ATS (flight supervision) unit;
- coordinating of the air traffic issues with units provided the aerodrome control and crew pre-flight information service (briefing).

Cooperation of the head, area, and regional centers of the Unified ATM System with the search-and-rescue divisions shall be maintained with the aim of the emergency notification.

Planning and coordinating of the airspace use for search-and-rescue operations shall be conducted in accordance with the state priorities.

Exchanging by the latest information on the aircraft in flight emergencies and other emergency situations occurred within the airspace shall be maintained in accordance with the table "Submission of a flight plan."

Cooperation of the head, area, and regional centers of the Unified ATM System with the air defense authorities shall be maintained with the aim of monitoring of compliance with requirements of Part 7 of the Federal Regulations.

The head, area, and regional centers of the Unified ATM System shall inform the air defense authorities about the plans (schedules) on the airspace use, as well as about the aircraft MVT through the information technologies provided by the Federal System for the RF airspace reconnaissance and monitoring.

Section Review

1. Tasks of the head center of the Unified ATM System.
2. Tasks of the area center of the Unified ATM System.
3. Tasks of the regional center of the Unified ATM System.
4. What authority is responsible for issuing the permission for the airspace use?
5. What are the procedures (sequences) of the restricted areas' use?
6. What planning shall be used for the airspace use?
7. What airspace classes are the follows: class A, class C, and class G?

References

1. Federal Rules on the Airspace Use, approved by the Enactment of the Government of the RF of 11.03.2010 No. 138 (FP IVP)
2. Federal Aviation Regulations on the Organization of Airspace Use Planning in the Russian Federation, approved by the Enactment of the Government of the RF of 16.01.2012 No. 6 (FAP OPIVP)

Chapter 2
Advanced Automated ATC Systems

2.1 General Information on New-Generation Automated ATC Systems

Over the recent decades, modern automated ATC systems have undergone considerable changes. New functions (TP, SYSCO, MTCD, MONA) have appeared, resulting in new opportunities for the ATC controllers, which were previously unavailable. The most significant change affecting the automated ATC systems was, most probably, the appearance of such option as trajectory prediction (TP). This function allows to provide the calculation of the so-called 4D trajectory, i.e., the forecast four-dimensional trajectory of the movement for each aircraft, the three dimensions being the special position (latitude, longitude, and height) and time being the fourth dimension. The 4D trajectory is described as an aggregate of 4D points and determines the A/C movement though space and time. The appearance of the 4D trajectories has made it possible to develop the function of medium-term conflict detection (MTCD), i.e., detecting the medium-term conflict situations on the projection horizon of 0–60 min and, in necessary, on a bigger range. Besides, the 4D trajectory allows automated ATC systems to be used for the automated monitoring on maintaining the trajectory, set by the ATC controller. The monitoring is realized by the MONA function (monitoring aids), which provides conformity monitoring and reminding. The main purpose of the MTCD function is to schedule the non-conflict movement, i.e., to schedule the non-conflict 4D trajectories of the A/C movement. The main purpose of the MONA function is to provide the automated monitoring on maintaining the scheduled 4D trajectories and to remind the ATC controllers of conducting certain actions. The MTCD and MONA functions provide absolutely new possibilities for the ATC controller, which have not been available with the automated ACT systems of the previous generations, and significantly change the concept of air traffic control. Now, the ATC controller is not only able to determine the conflict situations several minutes before they may occur, and to eliminate them, as it was already realized in the automated ATC systems of the previous generations. Now the ATC controller is able to reveal the possible conflicts considerably in advance and to schedule a

© Springer Nature Singapore Pte Ltd. 2020
Bestugin A.R. et al., *Air Traffic Control Automated Systems*,
Springer Aerospace Technology, https://doi.org/10.1007/978-981-13-9386-0_2

non-conflict trajectory, while the automated ATC system monitors the flight crew on following this non-conflict schedule.

The new functions demand the ATC controller to enter additional data inputs into the automated ATC system, such as entering the specified altitude and the altitude of transferring control from one ATC control area to the adjacent. However, despite the additional data inputs, these new functions extend the ATC controller's capabilities, thus increasing the airspace capacities of the control sectors and improving the air traffic safety.

The system-supported coordination (SYSCO) function, i.e., the coordination of the A/C movement, allows the automation of the coordination process and of the A/C control transfer from one ATC unit to the next, as well as the automation of the procedures for coordinating the conditions of AC transfer to adjacent control unit and for the accepting the control over the A/C from the adjacent control unit. With high intensity of air traffic, this function allows to decrease the ATC controller's work load, caused by existing routine manual tasks, thus contributing to the increase of the air traffic control capability of the ATC controller and to the increase of the air traffic safety.

According to the EUROCONTROL estimation, the introduction of such functions as TP, MTCD, MONA, and SYSCO contributes to the increase of the air traffic capacities of the control sectors by 15% in the upper control area and by 10% in the lower control area as well as provides a two-times increase of the air traffic safety.

The modern automated ATC systems normally include as follows:

(1) ATC ISC;
(2) airspace use planning ISC;
(3) aerodrome control tower ISC (A/D TWR ISC);
(4) technical management and monitoring subsystem;
(5) voice communication subsystem;
(6) data transmission ISC;
(7) ISC of meteorological information provision (ISC "METEO");
(8) reference information ISC;
(9) integrated systems simulator (ISS);
(10) universal time aids;
(11) a set of system operational manuals.

In the ATC ISC, the following processes shall be automated:

(1) receiving, processing, and displaying the information on the air traffic situation (the information on the observed conditions);
(2) receiving, processing, displaying, and distributing the information on the information on ATFM planning;
(3) receiving and processing of the meteorological information;
(4) analyzing the information on the current and forecast air traffic situation, based on the information on the observed conditions and on flight plan information. Automated process of detecting medium-term conflicts;

(5) displaying the information on the current and forecast air traffic situation, flight plan information, meteorological information, and the information on the airspace limitations;

(6) backup processing and displaying the information (in the "bypass" mode);

(7) recording and displaying the information, processed by the complex, including manual data input from the air traffic controllers' workstations;

(8) the information support for the calculation of the charges for air traffic services;

(9) the interaction between the airspace use planning ISC and A/D TWR ISC;

(10) the interface with the adjacent automated ATC system (in accordance with the OLDI protocol);

(11) the interface with the ministerial ISCs and automated management systems;

(12) the automated arrival and departure management (AMAN/DMAN) for the purpose of optimizing the sequence of aircraft takeoffs and landings;

(13) the technical management and control of the technical and software aids of the ATC ISC.

In the airspace use planning ISC, the following main processes shall be automated:

(1) collecting, processing, storing, and recording of the aeronautical and reference information, required for solving tasks related to planning the use of airspace;

(2) collecting/receiving, processing, storing, and recording of flight plans of scheduled flights, provided by the airline operators and by the Head Center of Unified ATM System;

(3) collecting, processing, storing, and recording of flight plans of unscheduled flights:

- ATS-route flights, provided by the Head Center of Unified ATM System;
- flight, performed partially or fully off the ATS routes, local flights, plans for use of airspace for other kinds of operations, involving the use of airspace, provided by the users of airspace and the interfacing ATS units.

(4) generating and recording the plan of airspace use of the Unified ATM System area;

(5) generating, transferring, and synchronizing the aerodrome map/chart for the aerodrome control tower and for the mobile ATS unit;

(6) collecting/receiving and processing of the information on the progress of fulfilling the plan of airspace use, providing the information on the plan amendments as well as the messages related to tactical planning of the airspace use to the ATM units and the airspace users of the Unified ATM System area;

(7) recording and displaying the information, processed by the ISC;

(8) the interface with the ATC ISC concerning the planned and aeronautical information, the information related to the limitations of airspace use as well as the information on actual fulfillment of the flight plans (use of airspace);

(9) the interface with the ATC ISC concerning the receiving of the current information on the observed conditions;

(10) the interface with the concerning the air defense units, the ministerial ISCs and automated management systems, the system of receiving ATFM plan data via

the Internet of the related ATM area concerning the ATFM plan, aeronautical and reference information;

(11) technical management and control of the operation of the complex.

In the A/D TWR ISC, the following main processes shall be automated:

(1) receiving, processing, and displaying the information on the observed conditions from the ATC ISC and from the aerodrome systems, means, and aids;

(2) receiving, processing, storing, displaying, and distributing the ATFM plan data;

(3) receiving, processing, and displaying of the meteorological information from the aerodrome automated information systems and the ISC "METEO";

(4) analyzing the information on the current and forecast air traffic situation;

(5) displaying the information on the current and forecast air traffic situation, flight plan information, meteorological information, and information on the airspace limitations;

(6) collecting, processing, and displaying the reference information;

(7) technical management and control of the operation;

(8) recording and displaying the information, processed by the ISC, including the input of the information from the air traffic controllers' and system engineer's workstations;

(9) the interaction with the adjacent aerodrome and airport automated systems, complexes, and aids;

(10) safety management of performing takeoffs and landings.

In the technical management and monitoring subsystem, the following main processes shall be automated:

(1) constant monitoring of the operational modes and technical conditions of the automated ATC system elements (means, aids, sources of information, data transmission channels, etc.);

(2) management of the operational modes, technical conditions, and configurations of the automated ATC system elements (means, aids, sources of information, etc.);

(3) generating and displaying the information on the functional control in the agreed form of displaying;

(4) data recording in the system log;

(5) diagnostic analyzing and testing of the automated ATC system aids in the automated mode;

(6) recording and displaying the information, processed by the technical management and monitoring subsystem, including the inputs from the workstations of the operating (engineering and technical maintenance) personnel;

(7) displaying the information on the air traffic situation;

(8) the option of selecting (switching on and off) the sources of radar information, incorporated into the system;

(9) the option of correcting (blanking) the processing areas of the radar fields depending on the sources of radar information and airway configurations;

(10) the provision of the unified time for the automated ATC system;

(11) administration, including database administration, and generation of reports;
(12) the option of exporting the technical management and control data to the external users;
(13) provision of reference information (in case of absence of automated reference information aids).

The voice communication subsystem shall provide the following:

(1) two-way communication in the very high frequencies (VHR) and ultra-high frequencies (UHF) ranges between the air traffic controllers of the units equipped with the ATC ISC and the flight crews in the telephone and loudspeaker modes, the two-way "ground-to-ground" communication in the VHR and high frequencies (HF) range with the moving objects as well as the option of listening to the "VOLMET" and "ATIS" radio channels at the workstations of air traffic controllers;
(2) internal operational telephone communication for the air traffic controllers and the system engineering and technical maintenance personnel;
(3) external operational telephone communication for the air traffic controllers and the engineering and technical maintenance personnel with the other aerodrome personnel;
(4) external operational telephone communication with the remote objects of interface with the automated ATC system (adjacent area control units, approach control units, airports and aerodromes of the control area, etc.);
(5) the option of the recordkeeping output of the voice information for the option of recording the information on the long-term storing unit.

The data transmission ISC must provide the exchange of voice and tele-code information (data) between the automated ATM system units with its peripheral objects, with external systems and integrated systems of ATC complexes, planning of airspace use/visual flight rules (VFR), automated information systems, etc. In the "METEO" ISC, the following processes shall be automated:

(1) receiving meteorological information, provided by the automated information systems—for example—from the "METEO SERVER" automated information system;
(2) generating and issuing the meteorological reports for the airspace use planning ISC, ATC ISC, and the A/D TWR ISC;
(3) displaying meteorological information on the autonomous displaying aids for the ATC personnel;
(4) generating notices on actual or forecast hazardous weather conditions which may affect the safety of flight operations.

In the reference information ISC, the processes of information input, storage, correction, and display of reference information for the needs of the operational personnel (air traffic controllers and the engineering and technical maintenance personnel) of the automated ATC system shall be automated.

In the integrated systems simulator (ISS), the algorithms for solving the following main tasks shall be automated:

(1) the simulation of performing the main technological processes of planning the use of airspace and the air traffic control, applied at the workstations of the area control center, aerodrome control unit, and aerodrome control tower;

(2) provision of the training of the engineering and technical maintenance personnel for the awareness of the operational procedures for the use of technical aids of the automated ATC systems.

2.2   Basic Functions of the Air Traffic Control in the Automated ATC Area Control and Aerodrome Control Systems

2.2.1   Purpose of ATC ISC

The general purpose of the implementation of the ATC ISC is the provision of assistance for the ATC personnel for performing their air traffic control and management tasks in order to achieve the following aims:

- the provision of efficient, safe, and quality services to the users of the airspace in the control areas within the responsibility of the ATC control area and in the airline hub control area within the responsibility of the consolidated control area of the Unified ATM System;
- the increase of the ATC capacity of the main aerodromes of the airline hub control area and the elements of the airspace;
- the implementation of modern ATC procedures;
- the automated informational interface between the ATM units and the users of airspace and the air defense units of the relevant Unified ATM System area.

In order to fulfill these functions, the ATC ISC shall provide the automation of the current planning of the airspace use and the automated air traffic control in the area within the responsibility of the ATC area control unit and the aerodrome area, including the following:

- automated collecting, processing, merging, and displaying the information from the primary and secondary radars (aerodrome and airways radars), from the automatic direction finders and the ADS for the display of the current air traffic situation to the air traffic controllers;
- automated collecting, distribution, and presentation of the FPL data, including the data exchange between the adjacent Unified ATM System units and the units responsible for the airspace use planning;
- collecting, distribution, and presentation of the aeronautical and meteorological information;
- recordkeeping and storing the data with the option of data displaying;
- provision of training for the ATC personnel.

2.2.2 Main Functions of the Information Processing for the ATC Provision

The implementation of the main (basic) functional tasks by means of the ATC ISC in relation to sectors of the area control units, aerodrome control units, and local control units includes the following:

- processing of the information and the option of solving the functional tasks of current ATFM planning and air traffic control;
- receiving and processing of the formalized reports in accordance with the ICAO Standards and the special formalized reports on ATC;
- automatic aircraft monitoring in accordance with the information on the observed conditions;
- continuous monitoring of the aircraft trajectory within the radar and ADS coverage area;
- processing of the flight plans (active and passive);
- receiving, processing, and displaying the information, obtained from the conditions observing aids, including primary surveillance radar (PSR) and/or secondary surveillance radar (SSR), automatic direction finder (ADF) (combined with radar and autonomous ones) and the ADS;
- the interface with the airspace use planning ISC as part of the automated ATM system providing the ATFM plan and aeronautical information, the information on the airspace use limitations as well as the information on the actual performance of the flight plans;
- the interface with the ISC "METEO" of the automated ATM system;
- the interface with the unified time aids of the automated ATM system;
- the interface with the adjacent automated ATC systems (in accordance with the OLDI protocol), the adjacent area control units of the Unified ATM System, the Head Center of Unified ATM System, the ministerial ISCs and automated management systems, the switching center of aerodrome ground-based system of data transmission and telegraph communication, the integrated system of the software and technical aids of the information technical interaction system, Russian Federation air traffic information collecting system, and the ADS-B aids (Technology 1090ES and VDL 4).

Processing of the information on the observed conditions of the current and forecast air traffic situation includes:

- receiving, processing, and displaying the information on the observed conditions of the current and air traffic situation from the PSR, SSR (including Mode S), SSR (with additional radar information on the aircraft state registration identification system, if available), the ADS (1090ES and VDL 4), including the following:
  - combining (tertiary processing) the information on the observed conditions from all the information sources;

- automatic filtering of the combined information on the observed conditions in respect to the information sources: independent and dependent surveillance;
- automatic input of the aircraft coordination points (plots) into the aircraft monitoring (trajectory filter), generating the smoothed estimates of the aircraft coordinated position (tracks);
- automatic linking of the information on the current flight plan to the track for the aircraft, equipped with ATC transponders and/or ADS-B, in case of the aircraft entering a certain airspace area;
- manual linking of the information on the current flight plan to the track for the aircraft, not equipped with the ATC transponders;
- performing the trajectory calculations concerning the aircraft movement parameters (ground speed, ground track angle, vertical speed, vector of the aircraft extrapolated position);
- automatic conversion of the aircraft flight radio altitude, obtained from the aircraft into the aerodrome runway elevation (QFE) or to the mean sea level (QNH).

- automatic discharge of the automatic aircraft monitoring and the aircraft release from the control in case the aircraft leaves the control area;
- automatic notification of the air traffic controllers on gaps and/or non-receiving of the coordination and/or additional information from the radar complex and/or ADS-B. In case of non-receiving the information on the aircraft from the radar complex during the N-number [varied system parameter (VSP)] of observations in succession, the automatic transfer of the last-received information on the aircraft into the loss list and the switching to the aircraft monitoring in accordance with the flight plan with the option of manual correction of the aircraft track position.
- automatic recovery of the automatic monitoring and displaying the monitoring log of the aircraft, provided with the flight plan and equipped with ATC transponders, manual recovery of the automatic monitoring of the aircraft from the loss list in case of obtaining the information on the aircraft from the radar complex.
- displaying the following information concerning all the aircraft, located in the area of responsibility of the ATC ISC and in the buffer area, on the ASD for all the ATC personnel:

  - current and previous (in accordance with the observation cycles) coordinates of the aircraft position;
  - additional information obtained from the aircraft, equipped with the ATC transponders (ATC transponder code, current altitude, and other parameters, obtained from the aircraft systems).
  - Automatic finding, based on the information, obtained from the observed conditions and from the air traffic controllers, and the displaying of the notifications:
  - on forecast of the separation minima transgression;
  - on the separation minima transgression;
  - on the forecast and the fact of the aircraft descent below the minimum safe altitude (MSA);
  - on special flight situations in case of receiving the emergency SSR codes and the code of 2000 by the complex.

Processing of direction-finding radio bearing information from ADF includes:

- receiving, generating, and displaying up to three radio bearing lines of the adjacent ADFs on the air situation display (ASD) of the air traffic controllers' workstations and the ensuring these ADFs' operation on the same frequency channel;
- the option to provide aircraft radio bearing on the tree frequency channels;
- displaying radio bearing lines of the adjacent ADFs no longer than 1 s after the aircraft contact on the ASD of the air traffic controllers' workstations and the termination of displaying the radio bearing no longer than 1 s after the contact has been finished.

Processing of ATFM information includes:

- receiving the daily plans of airspace use from the airspace use planning ISC, printing the daily plans of airspace use arranged by different criteria (on the traffic controllers' requests);
- processing and displaying the lists of repetitive flight plans (RPLs) and filed flight plans (FPLs);
- automatic (obtained from OLDI notifications) and manual activation of flight plans;
- automatic correction of ATFM plan information obtained from automatic monitoring from the surveillance system;
- correction of ATFM information in accordance with the received and/or manually entered ATC notifications;
- automatic calculation of the 4-D trajectories of the aircraft flight based on the data from the flight plan, corrected in accordance with the observation system data depending on the flight type, airspace structure, and flight management with regard to aircraft performance parameters and, the wind and the temperature of the airspace in the area, serviced by the ATC ISC;
- automatic calculation and displaying of the planned route with the turning way-points, ETO and flight level, and—for the starting and final enroute points—the aircraft call sign on the ASD on request;
- displaying minute intervals of the flight time by flags along the flight route based on the flight plan information;
- automatic monitoring over the turning waypoints passage and the control transfer points;
- manual switching to the automatic monitoring for the tracks, chosen by the air traffic controller for the aircraft based on the performed calculation of the 4-D flight trajectories in case of loss of the radar/coordinate information or beyond the area of radar coverage ("track by plan");
- automatic and manual distribution of ATFM information among the air traffic controllers' workstations equipped with the complex of aids of ATC automation in accordance with the calculated aircraft flight trajectory;
- linking of the ATFM plan information to the information on the observed conditions based on the code number of the secondary radar (SSR code) and the ICAO 24-bit aircraft code (for the ADS-B information);

- solving the tasks related to the system flight plan correction, including the assignment of a new call sign;
- solving the tasks related to assigning codes of the ATC transponders (SSR codes):
  - automatic monitoring over the use of the SSR codes in the RBS mode, over the presence of SSR binary codes in the in the RBS mode and the displaying of the respective annunciation for the air traffic controllers;
  - storing of one or several specified SSR codes;
  - automatic protection of the SSR codes from the possibility of the reuse;
  - analyzing the presence of the conflict codes, supertransit codes, codes which are forbidden for use;
  - automatic provision of the unoccupied code from the database in case of the plan activation by the entry points where the code change shall be performed;
  - automatic monitoring and displaying of memo messages related to the SSR code change.
- solving the tasks related to coordination and control transferring/accepting interface between the adjacent ATS units:
  - automated process of control transferring/accepting interface between the adjacent sectors of the system of aerodrome control units, area control units, remote sectors, and the airline hub control units;
  - manual transferring/accepting of the aircraft control;
  - manual releasing of the aircraft from control.
- automatic monitoring and message generation:
  - on control transfer;
  - on control acceptance;
  - on the point of ACT transfer to the adjacent control unit, which is not equipped with the instrumental interaction with the ISC in accordance with the OLDI standard.
- automatic monitoring:
  - in case the aircraft is approved for the RVSM flights (in case of RVSM mode implementation);
  - in case the aircraft is approved for the RNAV flights (in case of RNAV routes implementation);
  - in case for the approval for flights with the 8.33 kHz frequency grid.

Processing of the formalized ATC messages includes:

- receiving, logical and semantic processing, and displaying the formalized ATC messages in accordance with the document entitled "Table of messages for aircraft movement in the Russian Federation";
- automatic or manual correction of the flight plans based on the received messages;

- generating (automatic and manual) of the formalized ATC messages, their input and transfer via the aerodrome ground-based system of data transmission and telegraph communication from the air traffic controllers' workstations;
- automatic and manual forwarding of the received formalized ATC messages between the sectors within the Unified ATM System control area;
- recordkeeping of the incoming and outgoing messages;
- recordkeeping of all the correction messages concerning the exact flight plan and all the corrections (automatic and manually entered) related to the exact flight plan, both active and passive;
- ATC message distribution in accordance with the document entitled "Table of Civil Aviation messages" and the ICAO Standards.

Processing of meteorological information includes:

- receiving, processing, and displaying the meteorological information, including:
  - routine and special meteorological reports on actual weather, including the trend forecast, in the METAR/SPECI format for the aerodromes of the Unified ATM System control area, the aerodromes of arrival, and alternate aerodromes;
  - current data on aerodrome pressure and for other points of the Unified ATM System control area;
  - forecasts with the amendments in the TAF format for the aerodromes of the Unified ATM System control area, the aerodromes of arrival, and alternate aerodromes;
  - the forecasts for takeoff and landings;
  - local routine and special meteorological reports, including the trend forecast, in the MET REPORT/SPECIAL format for the main aerodromes of the airline hub;
  - actual runway condition information for the main aerodromes of the airline hub;
  - warnings for the main aerodromes of the airline hub;
  - wind shear warnings and alerts main aerodromes of the airline hub;
  - radar meteorological information on hazardous weather phenomena which may affect the aviation safety, types of clouds, clouds direction and speed, precipitation intensity, visibility, etc., for the Unified ATM System control area and the adjacent Unified ATM System control areas;
  - integration of the radar, satellite, radio bearing information and the observation data obtained from the meteorological offices on hazardous weather phenomena which may affect the aviation safety, related to cloud situation;
  - information on actual and/or expected enroute weather condition in the SIGMET and AIRMET code format for the airspace of the Unified ATM System control area and the adjacent Unified ATM System control areas;
  - area forecasts with amendments in the GAMET code format;
  - special reports from the aircraft onboard systems for the consolidated control area;

- data on the atmospheric layers in a regular grid for the airline hubs of the regular operations grid in the digital format for flight levels (to perform the calculations for flights planning and re-planning), including:
- forecasts for upper winds, upper-air temperature, and humidity;
- direction, speed, and flight level of maximum wind;
- flight level of tropopause;
- temperature of tropopause and geopotential altitude of flight levels;
- flight level forecasts of significant weather phenomena (turbulence, squall, icing, jet streams, cumulonimbus clouds associated with thunderstorms, etc.);
- enroute weather forecasts;
- wind data for altitudes of 30–100 m and for the circle altitude for the main aerodromes;
- ATIS information index for the main aerodromes of the serviced control area;
- information on launching, current position, and estimated shift of the radiosonde balloons for the atmosphere vertical sensing stations;
- advisory information on pre-eruption volcanic activity and/or a volcanic eruption;
- information on volcanic ash cloud and its trajectory for which the SIGMET information has not been issued;
- information on release of radioactive materials into the atmosphere;
- information on current and forecast weather for the state aviation aerodromes (on its receiving);
- on request: any meteorological information to meet the requirements of the airborne aircraft.

- receiving, distributing, and displaying (both on request and in a mandatory mode) the meteorological information on the workstations of air traffic controllers and the top management stuff;
- taking into account the meteorological data when solving tasks related to flight planning and re-planning as well as other system ATC tasks;
- mandatory displaying of the alerts/warnings on the workstations of air traffic controllers and the top management stuff in case the ATC ISC receives the associated meteorological information;
- automatic updating of the meteorological information on the workstations of air traffic controllers and the top management stuff in case the ATC ISC receives the associated meteorological information;
- automatic monitoring over the actual weather conditions to meet the requirements established for each type of the departing and arriving aircraft and for the respective PICs;
- automatic detection of the possibility for the aircraft to enter the area of hazardous weather phenomena, based on the observation information and the meteorological information, and generation of warnings for the air traffic controllers.

Processing of the information on airspace use limitations includes:

- displaying (on air traffic controllers' requests), amending, cancelling, storing, copying, and printing of the plans of airspace use, stored in the ATC ISC;
- generating and displaying the areas of airspace limitations on the ASD of the workstations of air traffic controllers;
- automatic and/or manual activation of the airspace limitations;
- automatic detection of the possibility for the aircraft to enter the area of airspace limitations, based on the observation information, and generation of the respective warning.

Alongside the above-mentioned functions, the ATC ISC can perform other basic functions:

- related to the special features of the ATC sector, located beyond the airways;
- backup processing and displaying of the observation information ("bypass" mode);
- recordkeeping and displaying of the information, processed by the ATC ISC including the input of the information from the air traffic controllers' workstations;
- interface with the A/D TWR ISC;
- interface with the adjacent automated ATC systems (in accordance with the OLDI protocol);
- interface with the ministerial ISCs and automated management systems;
- entering, storing, and updating the information on the conditions of communication aids, and radio communication equipment, aerodromes, reference information, information on aeronautical infrastructure of the controlled airspace and the ATC capacity of the elements of the ATM system;
- informational support for calculating the aeronautical services charges.

Additional functions of the ATC automation include:

- additional analysis of the information on current and forecast air traffic situation, based on the observation information and on the flight plans, including:

  - automatic monitoring of the values of the deviations from the designated flight trajectories;
  - recording of the deviations from the designated flight trajectories, including the recording of the aircraft flight number;

- monitoring over the trajectory maintaining and the generation of the warnings and memo messages (MONA function:
  Warnings:

  - on the deviations of the aircraft position from the current flight plan (regarding the longitudinal, vertical and lateral position);
  - on the deviations of the performed maneuvering from the designated;
  - on failure to meet the required control acceptance/transfer conditions;
  - on failure to maintain the climb/descent mode;

Memo messages:

- on changing the frequency, required for the control sector;
- on the necessity to switch to the manual amendment;
- on the planned maneuvering;
- on reaching the designated waypoints points.

- detecting the "medium-term" (within the time forecast period of 0–20 min) conflicts between two aircraft, detecting the forecast transgression into the special-use airspace, detecting the forecast descent below the lowest allowed flight level. These tasks shall be solved taking into account the data on the calculated 4-D aircraft trajectory, and shall be accompanied by displaying the associated warning and additional information on the conflict (in form of special windows for the detailed analysis of potential conflicts) for the air traffic controller;
- automated interface with the adjacent area control units and aerodrome control tower units for the purpose of coordinating the control acceptance/transfer conditions;
- automated ATFM of the approaching and departing aircraft in order to optimize the arrivals (the "AMAN" function) and departures (the "DMAN" function) for the main aerodromes of the airline hubs;
- automated estimation of the aircraft trajectory, ensuring the transition from the standard route to the alternate route as well as the backward transition to the standard route alongside with fulfilling the following instruction:

  - elastic vector;
  - direct route (the "DIRECT" function);
  - offset route (the "OFFSET" function);
  - instructing on heading alongside with instructing on the time period for maintaining this heading;
  - flight in the holding areas alongside with instructing on time for starting the approach.

- transferring of the radar surveillance information to the integrated ADS-B system for further transfer of this information on board the aircraft (the "TIS-B" function);
- collecting the statistic data on deviations from the advised flight trajectories;
- option of forming the information archives for the data related to the trajectory deviations in the established formats in order to transfer these data further to the external systems for further processing.

2.2.3 Displaying the Information on Air Traffic Controllers' Workstations

Structure of Human–Machine Interface (HMI)

This section dwells on the principles of organizing the interface in the automated ATC system as well as its functions, on the possibilities of its application, implemented in the JSC "VNIIRA" newly developed facilities.

Main Principles of Organizing the HMI

The implementation of the HMI in the Russian-built automated ATC systems is provided on the basis of the EUROCONTROL recommendations, aimed to ensure the safe, orderly air traffic flow of high intensity for the EU states while maintaining (and, possibly, increasing) the flight safety alongside with decreasing the workload on the air traffic controllers, taking into the account the specific technologies of the air traffic flow management in the Russian Federation, where the Cyrillic alphabet letters are used instead of the Latin alphabets.

The accepted approach allows to:

- decrease the workload on the air traffic controller's running memory and to use the ASD with the utmost efficiency;
- individually choose the system's interactive functions available;
- inform the air traffic controller on the results of the control panel inputs and ensuring the protection from making the wrong inputs (by generating the so-called diagnostic messages);
- individually adjust the displaying parameters at each workstation (such as the size of the dialog boxes, digits, linking lines, etc.) in accordance with the individual needs, thus improving the information perception and inputs management.

Technical Aids and Forms of HMI Organization

Based on the experience obtained while using the automated systems' and "SINTEZ" ATC aids' unified newly developed facilities, solid experience has been obtained in developing HMI system, which has been practically assessed and gained positive feedback from the air traffic controllers and which is designed on the basis of using the multi-window graphic interface for displaying the air traffic on the raster format display unit in form of coordinate symbols of aircraft, aircraft monitoring logs, various table-arranged lists, context menus, and dialog boxes with the option of their direct application for the purpose of performing the function calls or obtaining reference information.

2.2.4 Aids of Information Input

The main devices for providing the interface with the system and receiving access to all the functions, initialized through the objects identified on the screen and presenting the data, are currently a mouse and a keyboard. Each key of the mouse has a corresponding special type of input. The left key is for data input, the middle key (the scroll wheel) is for special functions (mainly graphical) with the scroll wheel function used for scrolling the lists, changing the image scale, and the right key is for calling out additional information and context menus. The keyboard is used for prompt entering of alphanumerical data into the system. In some cases, it can be used instead of the mouse to address the objects on the screen and to perform control operations.

It is recommended to have several means available to address the same function, including:

- the mouse;
- the keyboard functional keys;
- the keyboard hot keys (combinations of two or more keys).

2.2.5 Contents and Form of Displaying the Information on the Controller's Automated ATC System Workstations

On the indicator screen, there is the main window named the ASD with subsidiary windows in front—different dialog windows called out by addressing the control elements in the main window and independent windows called out while launching independent applications of the operation system by clicking on the window menu on the pop-up panel at the bottom of the screen. The ASD window is the most significant means of receiving operational information on the current and forecast airspace configuration, providing the display of all the related static and dynamic information on the background of the mapping information. Secondary basic windows can be continuously displayed on the ASD window, those including system data window, display control window, meteorological window, window of runway state and working mode, etc.

When addressing the control elements in those windows, different dialog windows can be displayed (user forms) which can be located in the appropriate area of the screen and are reset after finishing the interface. Here and below are the examples of human–machine interface elements implemented by VNIIRA in the automated ATC system "SINTEZ."

Figure 2.1 shows an example of displaying the information on the workstation of the area automated ATC system controller of Khabarovsk ATC regional center.

System Data In the system data window (see Fig. 2.2), the most significant data are presented: current date, time, designation, and operating mode of the workstation and other data in accordance with the Customer's request.

Mapping data The mapping data are static and represent a map of a given scale with elements of airspace structure and control area shown on it. The mapping data include:

- borders of the aerodrome area and highlighted control area;
- aerodrome approach/exit corridors;
- areas of constant limitations of airspace use;
- designated reporting points, waypoints, and radio fixes.

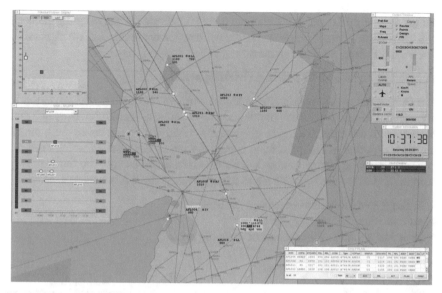

Fig. 2.1 Example of displaying the data on the workstation of an ATC controller. *Note* For detailed information on the contents and forms of displaying information in each particular system, it is recommended to use the related systems' manuals provided for ATC staff

Fig. 2.2 Example of the system data window

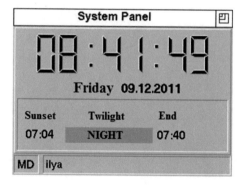

Other elements of cartography related to the aerodrome—aerodrome reference points (ARP) and their designations (of the relevant aerodrome and the airline hub, diverting and other relevant aerodromes), characteristic terrain points, turning points, boundaries of control acceptance/transfer, designation of artificial and natural obstacles, etc.—are chosen individually out of the set of mapping elements in accordance with the functions of the workstation. The exact contents of cartography are defined when forming a technical specification for the system. There is an option of input and display of individual maps at each workstation.

For map storage, the system database is used. There is an option of generating the maps individually (by the system administrator) with the help of Cartography

Editor. Operational control of switching on/off the display of the mapping elements and airspace structure is performed in the window shown in Fig. 2.3, which is called out from the display control window.

Prohibited Areas and Restrictions of Airspace Use

The data on the current restrictions of airspace use are shown in the ASD window as areas, in which airspace use is prohibited or restricted, as well as lines representing routes closed for flights. The areas of limitations of aerodrome use have a background, which is different in color from the common window background which does not cover the mapping elements and data on the aircraft movement. The limitations of aerodrome use are marked with a log, which displays the limitations parameters:

– indication of limitations of aerodrome use;
– altitude range in tens of meters;
– period of validity of limitations of aerodrome use.

Fig. 2.3 Display control window for mapping elements

(a)   **(b)**

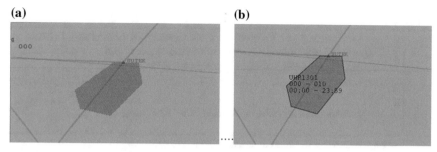

Fig. 2.4 Examples of displaying limitations of aerodrome use

The form of limitations of aerodrome use is displayed when placing the cursor onto the area of limitations of aerodrome use. Examples of displaying limitations of aerodrome use are shown in Fig. 2.4 (a—without a cursor pointing, b—with a cursor pointing).

Input of borders and parameters of temporary limitations is performed manually from the workstation of the flight dispatch supervisor.

Weather Hazards

Information on the weather hazards is displayed as areas whose borders are defined by the data entering from the system of meteorological support or entered manually. The weather hazards are displayed as colored polygonal contours. When placing the cursor onto the area of the weather hazard, the form of weather hazards with the characteristics of the hazard center is displayed. For getting attention, indication of urgent incoming meteorological messages on adverse weather conditions as well as warnings with further familiarization with the incoming message can appear in an individual window in the ASD window.

Examples of weather hazard areas are shown in Fig. 2.5.

Aircraft Coordinates Data The current coordinates' data on the aircraft, which is positioned within the visual range of the surveillance aids, are displayed as:

- coordinate symbols after the primary processing of the radar information ("plots");
- symbols of the monitored aircraft ("tracks") of various shapes, depending on the used surveillance aids, for example:

	– when providing support in accordance with data from the PSR only
	– when providing support in accordance with data from the SSR only
	– when providing support in accordance with combined data from the PSR and SSR

(continued)

(continued)

	– when providing support in accordance with extrapolated radar information (after stop of input of the radar information until the transfer of data from the aircraft monitoring log into the list of losses during 3–5 surveillance cycles, VSP)
●	– when providing support in accordance with the route data provided in the system flight plan
◇	– when monitoring a formation flight (multi-target monitoring)
■	– when generating warnings on possible flight safety hazards
▪	– when pointing on the track symbol or aircraft monitoring log with a cursor

The form of a track symbol can also show the characteristics of aircraft state registration. The coordinates' information of the automatic dependent surveillance system, transferred from the aircraft, which are equipped with the associated equipment, is displayed in an individual window. When a cursor is pointed on the track symbol or on the aircraft monitoring log, the track points turn green. The history of the aircraft track is displayed as a combination of points showing the positions of the track of the previous radar observations. The number of history points can be chosen when required. When the radar information ceases to be provided, the process of aircraft monitoring is maintained "out of memory" and, within a certain number of seconds, the number of points automatically decreases with each surveillance cycles until the displaying is stopped. Then the track symbol disappears, and the flight information transfers to the list of losses, in case the flight plan is unavailable. In case coordinate radar information disappears after "n" seconds, there are the following options (or a combination of them) for the aircraft in the area of the system operation:

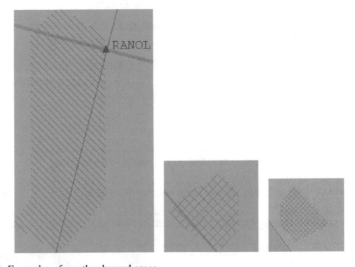

Fig. 2.5 Examples of weather hazard areas

- the aircraft track symbol is displayed continuously in the point of the last location with the characteristics of lost monitoring;
- in case the flight plan is available—automatic transfer of the track to the monitoring in accordance with the plan, with identification of the flight route, with an option of manual amendment of the aircraft position in accordance with the messages, received from the aircraft;
- in case the flight plan is unavailable—manual input of the track movement with the latest data on the flight bearing and speed, with an option of manual amendment of the aircraft position in accordance with the messages, received from the aircraft;
- automatic transfer of the last-received and planned (designated) aircraft data into the list of losses.

Radio Bearing Line Data from the automatic radio direction finder is to present azimuth to/from the aircraft and allow to find aircraft position and report magnetic bearing to the aerodrome control tower to the aircraft. These data are displayed as a radio bearing line and the values of direct and return azimuths for the aircraft in the middle of the line. Only one radio bearing line can be displayed at a time on the selected frequency. The bearing on the emergency frequency is always displayed by a red line, irrespective of whether the bearing is displayed or not. The values of the bearings are displayed according to the magnetic north. The bearing line is displayed during radio communication session with the aircraft crew.

Measuring Vector The purpose of the controller's tool "measuring vector" is to provide the option of monitoring the dynamics of the air traffic situation for the moving selected aircraft by means of measuring the azimuth and the distance between two points (geographical points or tracks in different combinations). The measuring vector is a line connecting one given point (or track) and another given point (or track). If the start and the end of the measuring vector are linked to the track symbol, such vector is called a "tracker." The data in the tracker log become dynamic and keep changing every time the radar data are updated. Together with the line, the tracker log is displayed with the information depending on the use of the measuring vector:

- The vector starting point is at a geographical point; however, the vector end points are linked to geographical points, so the azimuth to the second of the selected points and the distance between the two points shall be shown in the tracker log.
- The vector starting point is at a geographical point, and the vector end is at the tracker symbol; the tracker log shows the azimuth to the aircraft and the distance to it (in kilometers).
- The vector starting point is set at the tracker symbol, and the vector end is at a geographical point; the tracker log shows the azimuth to the point, the distance to it and the time of arrival at the given point, irrelevant of the aircraft heading, with estimation based on the current speed and the vector length.
- The vector starting point is set at the tracker symbol, and the vector end is at the other tracker symbol; the tracker log displays the azimuth to the second tracker and the distance to it.

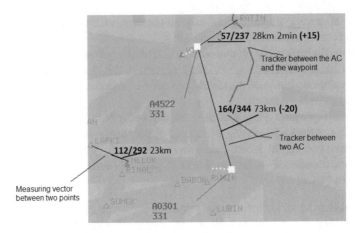

Fig. 2.6 Examples of displaying measuring vector and trackers

Examples of displaying the measuring vector and the trackers are shown in Fig. 2.6.

Any necessary number of measuring vectors and trackers can be selected and displayed at the same time in the ASD window. For these operations, only the mouth is used without any dialog windows. The forecast coordinates' data on the aircraft in the observation area are presented using extrapolated data of radar tracking in accordance with the selected parameter:

– after covering the prescribed distance (distance vector);
– after *n*-minutes of flight (speed vector).

The purpose of the distance vector is to evaluate the conditions of following the separation requirements and is an extrapolated leg of the aircraft track. The vector end shows the estimated position of the aircraft after covering the prescribed distance with the current speed and the current track angle, which are determined by the data of radar tracking.

The speed vector is an extrapolated leg of the aircraft track, where the vector end shows the estimated position of the aircraft after prescribed minutes of flight time with the current speed and the current track angle, which are determined by the data of radar tracking. The vector line contains index marks, which correspond to every minute of flight.

The displaying of the distance vector and the speed vector is alternate, i.e., selecting one vector type inhibits the displaying the other.

Displaying vectors for all the controlled aircraft in the sector provides an opportunity to estimate the forecast mutual aircraft positions. There is an option of operational selection of vector length values both for all the aircraft simultaneously (in the display control window) and individually for any controlled aircraft from the aircraft monitoring log (by addressing the associated field in the log).

For conflicting aircraft (with a near-collision tendency), the speed vectors (red) can turn on automatically with the value of 1 min (VSP).

The purpose of the aircraft monitoring logs is to present the main current data from the surveillance aids and the ATFM plan data used while controlling the flights. Besides, aircraft monitoring logs are used for system inputs in the automated ATC system, for example, for entering the prescribed altitude, heading, etc.

When the aircraft gets into the coverage area of the surveillance aids, the aircraft track is assumed to monitoring and is provided with an aircraft monitoring log.

The A/C monitoring log is connected to the track symbol by a link line. By default, the link line is at an angle of 135° to the A/C track angle. There is an option of switching on/off of the mode of automatic separation of the A/C monitoring logs in case of their overlap, an option of selecting the length of the link line in all A/C monitoring logs. Besides, the controller determines the angular position of all or of any selected A/C monitoring logs relative to the track line and places the selected aircraft monitoring log in an unspecified position on the screen. If the mode of automatic separation is on, the standard A/C monitoring logs go apart in case of overlap.

Types of the A/C Monitoring Logs' Displaying The A/C monitoring logs can be displayed in the following way: standard aircraft monitoring log, highlighted aircraft monitoring log, and extended aircraft monitoring log. Standard and highlighted aircraft monitoring logs are different in the contents for departing and transit aircraft and for aircraft performing landing at the aerodrome. Using the modern interface aids, the controller can perform control operations by addressing the necessary fields of a selected or extended aircraft monitoring log directly—for the most operational functions, or by a context menu called out by the right mouse key (RMK) (without pointing the cursor on any field of any parameter of the aircraft monitoring log).

The system has capabilities of operational control of the contents, forms, and elements of the information displayed in the A/C monitoring log. The number of aircraft monitoring logs displayed on the ASD screen shall correspond to the number of tracks. In case there is an A/C monitoring log with a message on a conflict condition and/or a distress code outside the screen, such information shall be displayed on a specific area of the screen.

Standard A/C Monitoring Log The purpose of the standard A/C monitoring log is to display the minimum necessary portion of information on all the A/C, positioned in the selected altitude layer (scheduled and non-scheduled). Standard A/C monitoring log allows the air traffic controller to present the most significant information continuously, and primarily—to estimate the A/C flight mode using the A/C monitoring log (entering/exiting the sector, instructed or not, approaching or departing, etc.). When the aircraft enters the area of radar coverage, the A/C track is assumed to monitoring and is provided with a standard A/C monitoring log. The standard A/C monitoring log also appears in case of manual linking of the A/C track to the input elements of the flight plan. The size of the log is determined by the flight configuration, i.e., whether there is a necessity to perform any commands (flight level change, etc.). The standard A/C monitoring log is semi-transparent and moves together with the track. Basically, the contents of the standard A/C monitoring log depend on whether the aircraft is equipped with a transponder or not. If the A/C is equipped with a transponder, the log contains the SSR code and the current altitude. If the aircraft is not equipped with a transponder (IFF or RBS), the log contains a special code (e.g., A0000) instead of the A/C call sign (an A/C identification). According to the recommendations of EUROCONTROL (and those implemented in VNIIRA ATC), the ASD window shows standard A/C monitoring logs in different colors, depending on the A/C flight "status" which determines the air traffic control capabilities for the A/C:

- "non-related aircraft"—the A/C has not yet entered or already left the control area, or the A/C monitoring log contains no FPL and has not been assumed to monitoring, the A/C monitoring log for such A/C is displayed in gray symbols;
- "activated aircraft"—after the FPL activation. The A/C monitoring log is displayed in dark blue symbols until the A/C is assumed to monitoring;
- "assumed aircraft"—the A/C (with or without an FPL) has been assumed to monitoring, i.e., has established radio communication with the air traffic controller of the sector; the A/C monitoring log is displayed in black symbols;
- "related aircraft"—the A/C is still in the sector, but has already been transferred to the ATC of the next sector, the A/C monitoring log is displayed in brown symbols (Fig. 2.7).

When describing aircraft monitoring logs, the following abbreviations are used:

ACID   aircraft identification;
AFL   actual flight level;
NS   next sector;
CFL   cleared flight level;
PEL   planned entry level;
TFL   transfer flight level.

When the A/C exits the observation area, the aircraft monitoring log stops being displayed.

The modern ATC systems also provide the capability for the controller to determine the contents of the standard and highlighted aircraft monitoring logs to some extent, depending on the current needs (except for the aircraft identification field).

Contents of Standard Aircraft Monitoring Log for Departing and Transit A/C

The standard A/C monitoring log contains the maximum of four information lines and an additional zero (upper) line and allows the controller to promptly estimate the aircraft flight mode.

ACID AFL	ACID NS AFL PEL RFL	ACID NS AFL CFL TFL	ACID NS AFL CFL TFL
Non-related AC	Activated AC	Assumed AC	Related AC

Fig. 2.7 Aircraft monitoring logs with different aircraft statuses. *Note* The specified colors are used in case of the gray background of the window

- Zero line:
  This line is used to display messages when coordinating and to display reminder messages and warnings/alerts.
- First line:
- If the plan elements have not been linked to the track:

  - SSR code (if there is no SSR code, A0000 is displayed).

- If the plan elements have been linked to the track, the aircraft identification is displayed:

  - radiotelephony call sign of the aircraft commander (formation commander) consisting of five digits (for state aviation aircraft when performing inner flights within the territory of the Russian Federation);
  - flight number (maximum seven symbols)—for civil aviation aircraft (when performing all types of flights); for state aviation aircraft when performing international flights.

  Then goes the sector designation—for the aircraft entering the sector this designation shows the previous sector, for the monitored A/C this designation shows the next sector (after sending the ACT message of activation). If the sector number is not entered, the special symbol is shown instead.

- Second line:

  - aircraft current flight level, AFL, in tens of meters, coming from the aircraft through the SSR channel. When the aircraft, equipped with a transponder, sends the altitude mode not as standard pressure (QNE), the letter "A" is displayed (preceding the altitude value if it is less than 10,000 m, and following the altitude value if it is more than 10,000 m). When monitoring an A/C by the PSR, the actual geometrical height measured by the radar (if the radar provides such capability) is displayed—as the letter "G" and digits indicating the height. If the height data are not updated, the altitude field shows dots. If there are no height data, the height field is empty;
  - arrow of the height change tendency (climb or descent) or an empty field when the system finds no tendency;
  - planned entry level, PEL (before entering the sector and being accepted into monitoring); in case there were no commands for level change before entering the sector, PEL becomes cleared flight level, CFL. CFL is not displayed if CFL = AFL ± 60 m (VSP);
  - ground speed if it is included for displaying by means of the display control window.

- Third line:

  - transfer flight level, TFL (is displayed after receiving the aircraft for monitoring only if it is different from the CFL), or requested flight level, RFL, if the A/C is in climb after takeoff;

  - exit point if it is needed and included for displaying by means of the display control window.

- Fourth line contains fields which are displayed only in cases the A/C is instructed:
  - flight heading (in accordance with the ADS-B data) or prescribed heading if entered by the controller;
  - current airspeed (in accordance with the ADS-B data) or recommended speed if entered by the controller (in km/h, Mach number or knots);
  - current vertical speed (in accordance with the ADS-B data) or recommended vertical speed of climb or descent (in m/s or ft/min);
  - instead of these data, one can choose to display temporarily the aircraft geographical coordinates.

Figure 2.8 shows the fields contained in a standard aircraft monitoring log for departing and transit A/C.

Figure 2.9 shows an example of a standard aircraft monitoring log.

When describing a standard aircraft monitoring log, the following abbreviations are used:

ACID aircraft identification;
CODE SSR code;
• Medium-term conflict detection (MTCD);
M/V/N status: military (M), or flight letter (A, B, V), or quantity of A/C in a formation during a formation flight (N);
NS designation of the current or next sector;

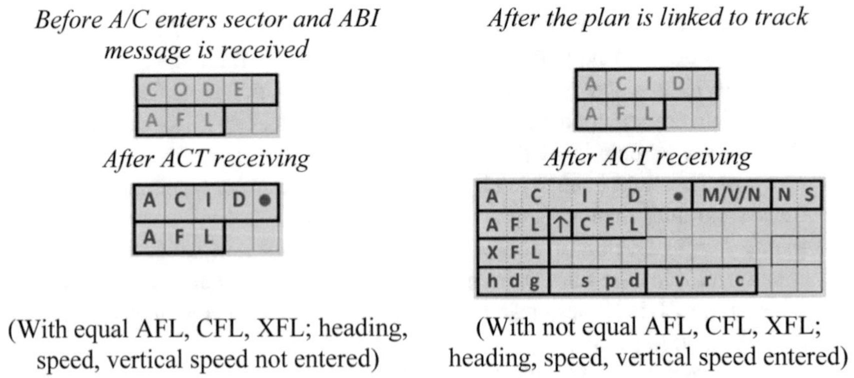
Before A/C enters sector and ABI message is received
C O D E
A F L
After the plan is linked to track
A C I D
A F L
After ACT receiving
A C I D •
A F L
After ACT receiving
A C I D • M/V/N N S
A F L ↑ C F L
X F L
h d g s p d v r c
(With equal AFL, CFL, XFL; heading, speed, vertical speed not entered)
(With not equal AFL, CFL, XFL; heading, speed, vertical speed entered)

Fig. 2.8 Fields in standard aircraft monitoring log for departing and transit A/C

Fig. 2.9 Example of standard aircraft monitoring log

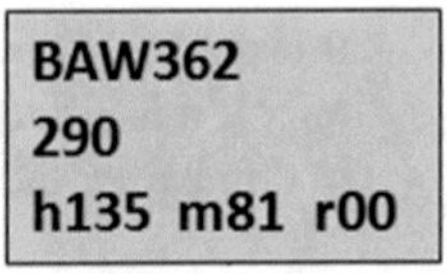
BAW362
290
h135 m81 r00

AFL↑   actual flight level (Mode C) and symbol of climb/descent;
CFL   cleared flight level (or planned entry level—PEL);
XFL   exit flight level (or transfer flight level—TFL);
hdg   prescribed heading or radar point;
spd   prescribed speed;
vrc   prescribed vertical speed of climb/descent.

Contents of a Standard A/C Monitoring Log on the ASD for Arriving Aircraft
Figure 2.10 shows an example of a standard A/C monitoring log for arriving A/C.

The standard aircraft monitoring log on the ASD for arriving aircraft includes:

- Zero line: the same as described above.
- First line:

  – aircraft identification suffix;
  – aircraft wake turbulence category (is entered by control panel operation);
  – sector designation (previous—before the start of monitoring, current—after the start of monitoring) or (for monitored aircraft) landing RWY (is entered by control panel operation).

- Second line:

  – actual flight level with datum level and change tendency;
  – cleared flight level, CFL, or selected type of approach—instead of "CFL" when type of approach is entered;
  – ground speed;
  – recommended landing time and delay time (with AMAN function engaged);
  – A/C type.

- Third line:

  – flight heading (in accordance with the ADS-B data) or prescribed heading if entered by the controller. This field can display not only the heading value, but also the abbreviated designation of the current element of the approach procedure;
  – indicated airspeed (in accordance with the ADS-B data) or recommended speed if entered by the controller.

Highlighted Aircraft Monitoring Log When placing the mouse cursor onto the standard aircraft monitoring log field, the track symbol, history track, extrapolated vector (if there is any), and link line turn green, and the log becomes highlighted.

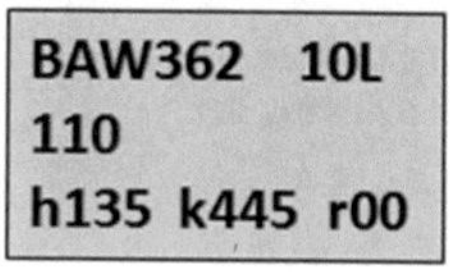

Fig. 2.10 Example of standard A/C monitoring log for arriving aircraft

BAW362   10L
110
h135 k445 r00

The purpose of the highlighted aircraft monitoring log is to display additional information and to address its fields to perform control operations.

Contents of a Highlighted Aircraft Monitoring Log for Departing and Transit Aircraft and Available Functions

The highlighted aircraft monitoring log contains all the fields of the standard log, i.e., AFL, CFL and TFL, even if their values are equal, airspeed, as well as the following data:

- in line 3—delay time (time aircraft shall be put on hold, calculated by AMAN);
- in line 4:

  - field designation, or flight heading (in accordance with the ADS-B data) or prescribed heading if entered by the controller;
  - field designation, or current airspeed (in accordance with the ADS-B data) or recommended speed if entered by the controller;
  - field designation, or current vertical speed of climb or descent (in accordance with the ADS-B data) or recommended vertical speed if entered by the controller.

Figure 2.11 shows the contents of the fields of a highlighted aircraft monitoring log, and Fig. 2.12 provides an example of a highlighted aircraft monitoring log.

When describing a highlighted aircraft monitoring log in addition to a standard log, the following abbreviations are used:

W   ground speed calculated by the system;
XPT   sector exit point (three symbols);
TYPE   aircraft type (in the ICAO format).

When using a highlighted aircraft monitoring log, the following operational functions are available:

- entering the prescribed flight level;
- entering the prescribed heading for a random geographical point from the current aircraft position (including the final turn);

Fig. 2.11 Contents of fields of highlighted aircraft monitoring log

A	C		I		D		•	M/V/N	N	S
A	F	L		C	F	L		W		
X	F	L	X	P	T		T	Y P E		
h	d g			s	p	d		v r c		

Fig. 2.12 Example of highlighted aircraft monitoring log

BAW362 **V3**
290 290 475
290 FK 45 H
h135 m81 r00

– entering the recommended airspeed;
– entering the vertical speed of climb/descent;
– entering/resetting the vector of predicted aircraft position;
– displaying the aircraft geographical coordinates;
– for arriving aircraft—calling out/removing the azimuth and distance to the THR.

Extended Aircraft Monitoring Log Extended aircraft monitoring log is called out by the controller to receive more information from the aircraft flight plan without calling out the flight plan itself or searching for the information in the lists. In operational conditions, the extended aircraft monitoring log can be used to send the ACT message to the next sector to activate the flight plan. The extended aircraft monitoring log can be placed near the highlighted aircraft monitoring log or in any available area of the screen.

Displaying of the Results of Automatic Air Traffic Situation Analysis To display a short-term conflict alert (STCA), the aircraft identification, track symbol, and history track points are highlighted (in yellow—for reaching the separation limits, in red—for near-collision, flying below minimum safe altitude, flying in areas of airspace use limitations or hazardous weather conditions).

Additionally, red speed vectors can be displayed automatically for conflicting aircraft after 1 min of flight.

Figure 2.13 shows the example of displaying STCA.

Figure 2.14 shows an example of an alert window.

The alert window always has the highest priority of displaying, i.e., it is always displayed above all the other windows, cannot be minimized, changed in size, or cleared from the screen. It can be operationally moved on the screen and placed in any convenient area. When the alert condition is gone, the corresponding line in the window is automatically removed from view. When there is no information for displaying, the window is automatically removed from view.

Fig. 2.13 Example of displaying STCA

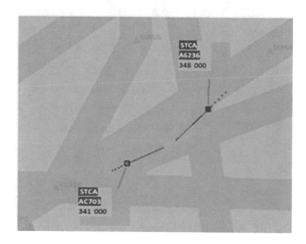

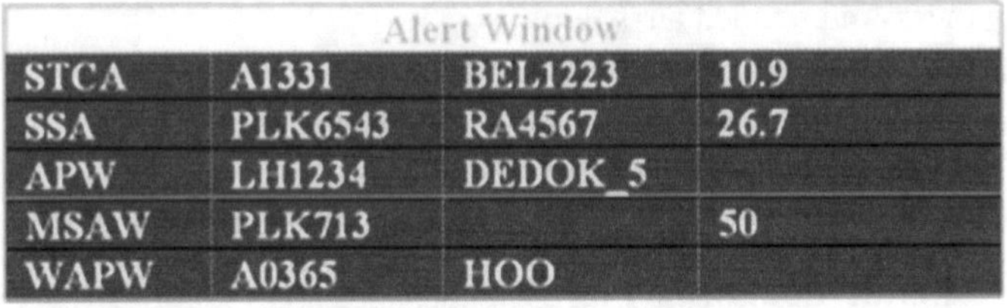

		Alert Window	
STCA	A1331	BEL1223	10.9
SSA	PLK6543	RA4567	26.7
APW	LH1234	DEDOK_5	
MSAW	PLK713		50
WAPW	A0365	HOO	

Fig. 2.14 Example of an alert window

Displaying Weather Information, RWY Condition, and Operational Mode
The window "RWY METEO" on the controller's workstation shows the actual meteorological data on the chosen RWY, the RWY condition, and the technical condition of the radio navigation means for the RWY. The window can have a smaller or a maximum size and display the minimum necessary or the whole contents of information by the operational choice. The window can be moved to any area of the screen. Figure 2.15 shows an example of actual weather conditions on an aerodrome RWY.

Displaying of an Air Traffic Situation Fragment (Zoom Window) The zoom window shows an ASD window fragment around a point on the ASD, selected by the controller, to monitor the selected air traffic situation (e.g., a potential conflict situation, a helicopter approach to a landing ground). The zoom window displays the selected area of the ASD completely, including cartography, areas with limitations,

Метео ВПП 1

ВПП		07 ПОЛОСА СВОБОДНА 25				ВКЛ	
Time	RWY cond. / Сост. ВПП		mm	%	B/A TDZ	MID	END
14:30	УПЛ. СНЕГ		10	25	0.50	0.50	0.40
QFE,мм	T,°C	Wind(2min),deg	mps	max,mps	Vis,m	RVR,m	H,m
742	-05	07 210	05	08	1200	1400	50
QFE,гПа	Td,°C	MID 230	04	05	1000	1100	H,m
998	-07	25 240	05	06	1100	1200	60
QNH,гПа	RH,%						
1028	80						
Пр.мин.,мм	Wind,deg	mps	max,mps	mps	mps	Cld N/Nh	Cld
760	240	05	06 ✈ 1,4 ✈ 7,1			ЗНЧ РЗБ	Cb

Phenomena/Явления
СНЕГ / SNOW
TREND/Прогноз на посадку
БИЗМ

| КГС | ДПРМ | БПРМ | ДВОР/ДМЕ | АТИС | A |

Fig. 2.15 Example of actual weather conditions on aerodrome RWY

areas of hazardous weather conditions, and aircraft monitoring logs. The window size can be changed both horizontally and vertically. The window can be removed from view.

Figure 2.16 shows an example of a zoom window for a random ASD window fragment.

Displaying of the Flight Plan During flights, displaying of ATFM plan data in lists and aircraft monitoring logs is performed based on the activated flight plans from the daily plan. The complete flight plan includes all the field in a single unified form recommended by ICAO for making flight plans, as well as specific additional fields based on the Customer's needs. For operational flight monitoring in a specific ATC segment and landing area, it is enough to use the plan elements in a simplified form including only the fields necessary for controlling the flights in these areas. The plan in a simplified form is called out by addressing the highlighted aircraft monitoring log.

Figure 2.17 shows an example of a flight plan window.
The flight plan window provides:

– operational entering of the flight plan elements data into the system and automatic linking of the flight plan data to the radar information when an aircraft monitoring log with the SSR code appears on the ASD;
– modification of the already entered plan;
– search of a plan using the aircraft identification from the previously entered flight plans;
– removal of the plan from the system.

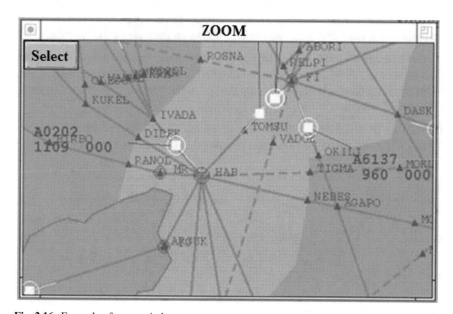

Fig. 2.16 Example of zoom window

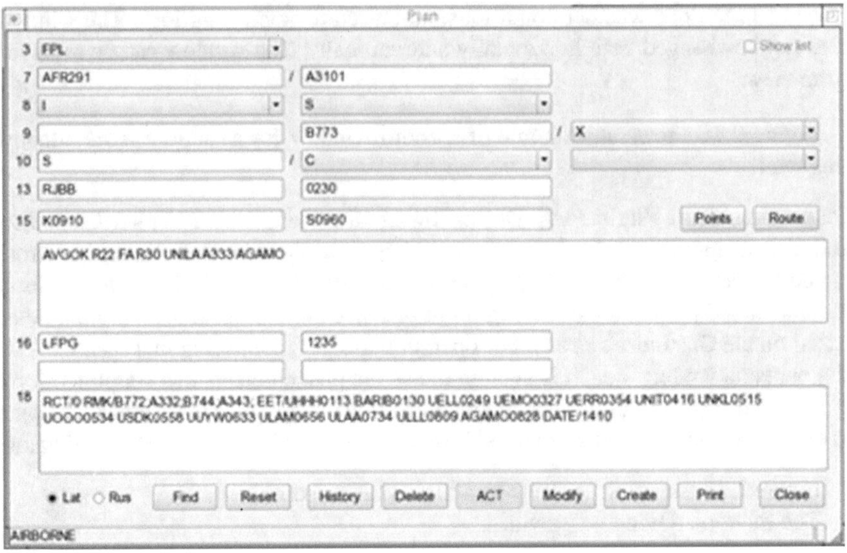

Fig. 2.17 Example of a flight plan window

Activation of flight plans can be performed after receiving a relevant message from an adjacent control unit through digital communication in an established protocol or manually using voice communication or radiotelephony communication.

Entry List (SIL LIST) The purpose of the entry list is to present data on the arriving and transit aircraft with forecast for the established time period in advance. This is for the aircraft, which are to enter the sector and allow forecasting of air traffic situation and estimating the sector load. Figure 2.18 shows an example of an entry list.

The information fields in the list lines are determined by the Customer based on their needs and ATFM plan data available. The lines are sorted by time of entry into the sector in relation to actual time (with the first aircraft to enter the sector displayed at the top of the list).

List of Controlled Aircraft The purpose of the list of controlled A/C (CONTROLLED AIRCRAFT) is to present additional aircraft data after it is taken under

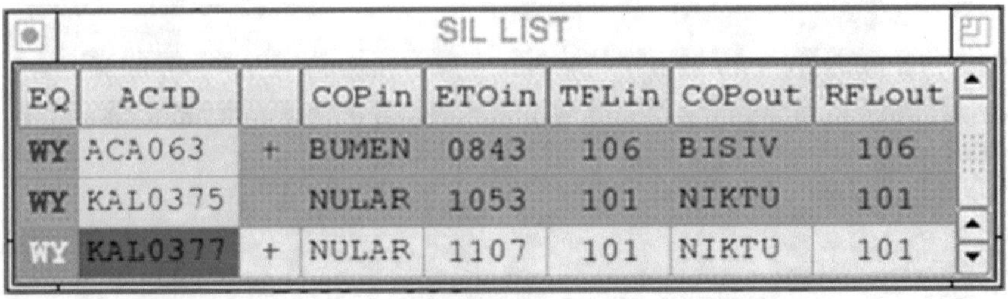

SIL LIST

EQ	ACID		COPin	ETOin	TFLin	COPout	RFLout
WY	ACA063	+	BUMEN	0843	106	BISIV	106
WY	KAL0375		NULAR	1053	101	NIKTU	101
WY	KAL0377	+	NULAR	1107	101	NIKTU	101

Fig. 2.18 Example of entry list

MTCD	ADEP	EQ	LIT	ACID	CODE	TYPE	CFL	COPout	ETO	TFL	RFL	DEST/RW	REM/WRN
	RKSI	WY		KAL037		B744/X	101	NIKTU	1156		101	KORD	
	RKSI	WY		KAL0376		B744/X	101	NIKTU	1241		101	KORD	

Fig. 2.19 Example of controlled aircraft list

control and to coordinate the conditions of transfer to the next adjacent sector. Figure 2.19 shows an example of a controlled aircraft list.

The list displays departing, transit, and arriving aircraft taken under control.

Hold List The purpose of the hold list window (HOLD LIST) is to present information on the aircraft sent to the holding area and is displayed in case there are such aircraft as a list where each line refers to a certain aircraft. Figure 2.20 shows an example of a hold list.

Coast List The purpose of a coast list (COAST LIST) is to save data on controlled aircraft in the sector in case the aircraft is lost from monitoring or in case there have been no linking of the plan to the track.

The contents of the list fields are determined by the Customer (e.g., the aircraft call sign (flight number), SSR code (registration number), aircraft type and weight category, current altitude and cleared flight level, point, altitude, and time of exit from the control area). Figure 2.21 shows an example of a coast list.

Transfer of an aircraft to the coast list is performed automatically. In case the radar data from aircraft with transponders start being received again, the data move from the coast list to the aircraft monitoring log automatically.

Control of the SSR Code Setting for Correctness

The system shall control the setting of SSR codes for correctness and generate a message on the necessity to change the SSR code in the following conditions:

LIT	ACID	CODE	AFL	CFL	COPout	CC	ALERT
	DFGHJHJ	30563		066			

Fig. 2.20 Example of hold list

LIT	ACID	CODE	AFL	CFL	COPout	CC
	KAL0377			101	NIKTU	TA

Fig. 2.21 Example of coast list

- The code is prohibited for use if the controller has entered an incorrect code by mistake.
- Vacant codes are currently unavailable.
- The system has two identical codes.
- If the ACT message contains an SSR code which cannot be used in the current area, the system offers a vacant code which will be displayed in the zero line of the controller's aircraft monitoring log to send to the aircraft.
- In case the aircraft sends the code 2000 (i.e., the code is not assigned).

These messages are displayed in the zero line of the aircraft monitoring log.

Transfer/Acceptance of Control (Without Transmitting Messages Through DDL)

When the aircraft comes to the transfer boundary (during a prescribed time before the estimated time of reaching the coordination point, usually 2 min), the system automatically switches on the mode of transfer/acceptance and generates a reminder message on the aircraft control transfer/assuming (on the change of the control sector frequency) to the controlling sector, in case the control transfer/acceptance procedure has not been initiated. The reminder messages are displayed in the zero line of the aircraft monitoring log prescribed time prior to reaching the control transfer boundary (the default value is 8 min, VSP). The message is removed after the established time interval of 30 s (VSP) after the boundary passing. The transferring controller shall instruct the aircraft to transfer to the adjacent ATC unit's controller by voice and then perform transfer operation on the control panel (by pressing the left mouse key (LKM) onto the sector designation in the aircraft monitoring log). Then a menu comes out (Fig. 2.22) with the highlighted TRANSFER function.

After transferring the communication and in case the transferred aircraft is still in the transferring sector of ATS, the aircraft monitoring log changes the aircraft color to "related" (amber). After the aircraft passes the sector boundary, the aircraft monitoring log becomes "non-related" (gray), which means the control has been transferred. When the aircraft leaves the controlled area and the plan is removed, the aircraft monitoring log displays in Fig. 2.23.

In case the transferring controller has not performed the TRANSFER operation before the aircraft passes the boundary of the area of ATFM plan data processing, the system automatically reminds the controller on the necessity to transfer the control by displaying the TRANSFER annunciation in the zero line (Fig. 2.24).

The message on the control transfer/acceptance is removed after entering the frequency change, or accepting the aircraft for communication, or when the message-

Fig. 2.22 Transfer of control by TRANSFER function

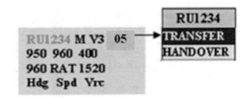

Fig. 2.23 View of aircraft monitoring log after A/C leaves control area

Fig. 2.24 Message on necessity to transfer control in zero line of aircraft monitoring log

generation conditions are gone. In case the transferring controller chooses to transfer the aircraft before the automatic transfer/acceptance boundary, the controller, having previously coordinated the transfer conditions through the voice communication system, instructs the aircraft to contact the next controller and addresses the sector field of the aircraft monitoring log, then selects the HANDOVER function and performs the above given procedure.

Assuming/Accepting Control When the aircraft contacts the accepting controller, the accepting controller shall address the sector field by the LMK. The menu with the selected ASSUME function comes out.

Figure 2.25 shows the aircraft monitoring log of the accepting controller when control is being accepting (the ASSUME function).

The controller addresses the ASSUME function by clicking the LMK, after that:

– the menu is cleared;
– the aircraft monitoring log becomes "controlled" and the log symbols become black;
– the sector designation changes for the sector to which the aircraft will be next transferred (for the next planned sector);
– the planned entry level PEL becomes the cleared level CFL;
– the TFL field shows the sector exit level in case it is different from the CFL;
– the line is cleared from the entry list;
– in the daily plan and planning lists, the line moves to the controlled aircraft list.

Fig. 2.25 Assuming control with ASSUME function

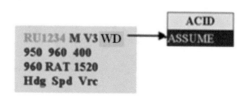

Display Control of the ASD

The display control of the ASD is provided by a special window, which is a set of control elements combined by their functions and providing prompt access to modifying the data display and location in the air data window. Figure 2.26 shows an example of a display control window.

When expanding the window, the set display parameters are indicated, including:

– the sector flight level range outside which tracks and aircraft monitoring logs of "non-related" aircraft are not displayed ("altitude filter"); in case the system restarts, the altitude filter sets at the maximum visibility automatically;
– the set scale of displaying;
– the set parameter of displaying extrapolated vectors for a prescribed number of flight minutes or for distance (for all the monitored controlled aircraft, including "turned-off" condition);
– the condition of turning on/off of displaying radio bearing lines, ADF frequency;

Fig. 2.26 Example of display control window

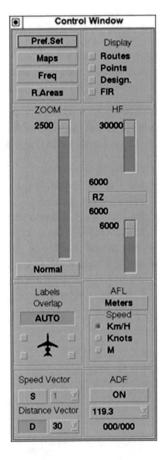

– the view of displayed data on the current altitude in the aircraft monitoring log (in feet or meters).

Addressing the control elements in the window, one can call out the related dialog windows and set the new parameters of displaying the above-mentioned data. Besides, one can call out the following dialog windows (menus):

– for choosing the number of dots in the track history, length of link lines, turning on/off the automatic clearing of aircraft monitoring logs, location of the aircraft monitoring log relative to track headings (in case the automatic clearing is off), turning on/off of distance marks relative to the selected point on the screen and their values;
– for selecting the cartography elements in the air data indicator window;
– for displaying the parameters of the set areas of limitations of airspace use (in case such function is available);
– for looking at the radio frequencies of the adjacent ATS sectors.

The display control window can be moved to any area of the screen, minimized to an icon, and expanded again. The settings can be saved as individual for the current user or set as default being optimal for the workstation. In the minimized view, the window stays on the screen and can be located the most appropriately. It can display the minimum necessary contents of data, e.g., limits of altitude filter, values of direct and reverse radio bearing during a communication session, etc.

Displaying Additional and Reference Data

These windows include windows which are called out and cleared by the user by addressing the related icon on the special panel (the same as the task bar in Windows OS). The contents of such windows depend on the purpose of the workstation and the tasks performed. They include the following examples:

• window of status and work modes of the data sources (radar, ADS-B, ADF), RWY condition;
• METEO window with meteorological data (hazardous weather conditions, METAR/TAF, SIGMET, AIRMET, GAMET, WIND, radioprobes' trajectory);
• window with data of coordination with adjacent automated planning means and ATC;
• window of input/clearing of data on obstacles on the RWY. It allows to input the data on the presence/absence of obstacles on the RWY according to their types (A/C movement, aerodrome means, etc.) which allows to control the RWY occupancy and automatically provide the related indication to ATC workstations;
• TLG window which contains a form to display incoming telegrams, a form to complete and to send telegrams;
• window of limitations of airspace use which provides displaying the current limitations of airspace use with indication of time of validity and flight level range. From the workstation, one can also perform amendment of limitations, input of new ones, and cancellation of existing ones;
• window to display and print out the statistics;

- window of reference data. While working in this window, one can operatively get reference data on the issue concerned, e.g., catalog of aviation equipment in different countries, tolerances of crosswind component and friction coefficient for A/C types, sunrise and sunset time by month, approach patterns and exit from control area, specific flight operations, abnormal operations guidelines, etc. From this window, one can also display operations manuals. The contents of the necessary data are determined by the Customer;
- system log with search and selection functions;
- other windows.

The functional windows of applications for special functions of ATC automation which are described next (such as MTCD, AMAN, DMAN) have specific significant purposes.

2.3 New Functions of the Modern ATC AS (TP, MONA, SYSCO, MTCD, AMAN, DMAN)

2.3.1 Prediction of 4-D Trajectory

General The term "trajectory" stands for plotting the aircraft movement in horizontal and vertical planes in the course of time. The term "trajectory prediction" ("TP") stands for the calculation of the 4-D trajectory, taking into account:

- tactical (current) flight plan;
- weather information;
- aircraft operational performance parameters;
- radar data on the aircraft movement;
- as well as the air traffic controllers' clearances.

The automatic calculation of the 4-D trajectories of the aircraft flights is conducted on the basis of the flight plan data, amended in accordance with the actual observation system data, airspace structure, and the ATFM, taking into account the aircraft operational performance parameters, the wind and temperature airspace, controlled by means of the automated ATC systems. The planned and tactical trajectories can be distinguished. The planned trajectory is a mid-term (from several hours prior to the beginning of the flight to one or two minutes prior to the current time) representation of the flight trajectory with the indication of the flight intent, described in the flight plan, taking into the account the limitations, defined by the ATC procedures. The planned trajectory is the basis, upon which the flight data are to be distributed item-by-item through the sectors, the aircraft is to intersect during the flight, the coordination among the sectors and among the ATC units is to be performed, the planning of the sector is performed, the mid-term conflicts are to detected as well as the deviations from the planned flight intent are to be checked. Immediately after the flight is started, the trajectory may be modified in accordance with the instructions of tactical planning and monitoring. The tactical trajectory ensures the short-term trajectory plotting in accordance with the most recent air traffic controller's clearances, issued for the aircraft without any speculation on the further air traffic controller's clearances to be provided. The tactical trajectory allows to reveal conflicts, which may have occurred if no further tactical instructions have been provided; as if the aircraft continued their flights with the current air traffic controller's clearances. This is useful for the relatively short-term prediction (e.g., of 5–10 min), after which further relevant tactical instructions are expected to be provided.

Interaction of TP function with the other functions

The interaction of the TP with other functions is shown in Fig. 2.27 and is described as follows [1]:

– flight data distribution (FDD)—the flight data are distributed among the workstations of the air traffic controllers in relation to the associated control sectors in accordance with the list of the sectors through which the trajectory runs;
– initial flight plan processing (IFPP)—at the stage of the flight planning, the trajectories are plotted and amended in accordance with the messages, received from the integrated system of airspace use planning automation and the flight data, manually entered by the system operators;
– SSR code assignment—the Mode A code is assigned on the basis of the planned route and the code distribution plan;
– flight data updating (FDU)—as soon as the flight plan is activated, the trajectory is amended in accordance with the tactical limitations (air traffic controllers' clearances and the instruction on sector planning), entered by the air traffic controllers;

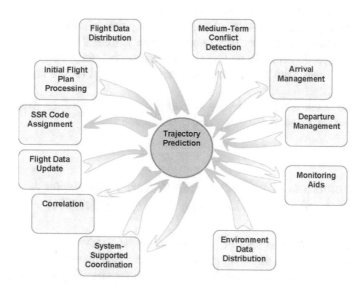

Fig. 2.27 TP function correlation with other functions

– correlation—the flight data are correlated with the associated system track based on the Mode A SSR code and the planned route;
– system-supported coordination—the coordination, which is performed automatically among the sectors and the external ATM units in accordance with the cross-sector control transfer conditions, contained in the trajectory; the trajectory is also amended in accordance with the cross-sector control transfer conditions received from the external ATM units;
– environment data distribution—information, related to the aeronautical elements (e.g., navigation aids, and ATS routes marking), which are mentioned in the flight plan messages. These data are used to open the planned route; aircraft operational performance parameters are amended in accordance with the weather forecast; the trajectory is modified with the limitations in accordance with the standard ATC procedures and agreements; the trajectory intersections with certain airspace elements (such as the special-use airspace and sectors) are determined;
– monitoring aids (MONA)—warnings on deviations and automatic memo messages. They are provided to the air traffic controller, based on the data on the route in the horizontal plane and the estimates of time, required for passing certain route waypoints, defined accordingly by the trajectory; as the actual flight proceeds, the trajectory is renewed in relation to the longitudinal and vertical measurements, which are determined on the basis of the correlated system track data;
– medium-term conflict detection (MTCD)—for planning and tactical support, aircraft conflicts are detected, where the calculated separation between the two aircraft, obtained from their positions on the trajectories in a certain moment of time is beyond certain limitations; similarly, aircraft conflicts with airspace and aircraft conflicts terrain are detected when certain criteria of the aircraft position go beyond the limits in relation to the airspace affected with the airspace use limitations or between the aircraft and the terrain obstacles;
– arrival management (AMAN)—the optimized sequence of arriving aircraft is generated together with the advisory information on additional time, required for the aircraft to spend in order to avoid the accumulation of traffic in the flow of arriving aircraft. Actually, the AMAN function can be also used for the arrival runway designation;
– departure management (DMAN)—the optimized sequence of the departing aircraft, also used for the departure runway designation.

2.3.2 Automatic Monitoring Over Prescribed Trajectory Maintaining and Reminder Messaging (MONA)

General Information on MONA The monitoring aids (MONA) are the instrumental aids, additional to the trajectory prediction, which assist the air traffic controller in monitoring all the controlled aircraft in order to detect deviations from the prescribed movement parameters in the system trajectories, to trigger the subsequent annunciation for the air traffic controller and to initiate the trajectory re-calculation [2].

The MONA includes:

- correlation monitoring: The MONA compares the current flight data with the system trajectory. In case any deviation is detected, MONA:
  - either automatically initiates the process of the trajectory re-calculation;
  - or generates the data to warn the air traffic controller.
- reminder messages: The MONA generates the memo messages in order to remind the air traffic controller of the actions which shall be performed.

As soon as the correlation monitoring aids detect the deviations of the actual aircraft position from the expected, depending on the type of the deviation detected, either the trajectory is to be automatically re-calculated or a "non-conformance" warning (NCW) is to be automatically displayed.

The "non-conformance" warnings (NCWs) shall have a lower priority than the SNET (the software aids for monitoring the flight safety) warnings; in other words, the NCWs shall not be displayed if the SNET warning is displayed. The NCWs shall be displayed for all the aircraft, including the state and experimental aviation aircraft. The memo messages are displayed in order to remind the air traffic controller about the actions related to the aircraft flight, which shall be performed. When the aircraft passes a certain "reminder" waypoint on its trajectory, the system shall trigger the displaying of the associated memo message. The "reminder waypoint" can be created automatically or as a result of a manual input.

The memo messages are displayed for the sector, with respect to which the aircraft can have one of the following statuses: activated, accepted, and relevant. The MONA receives system trajectory data, which provide the 4-D trajectories, plotted by means of the TP function.

Moreover, the MONA receives data on the current aircraft position and on its movement parameters (heading, speed, altitude, vertical speed, and their change trend). In order to provide high-quality functioning of the MONA, the radio coverage in the ATC area shall meet the EUROCONTROL requirements.

In the automated aerodrome ATC system of the Khabarovsk ACC, the following warnings can be generated:

- aircraft lateral deviation from the system trajectory;
- aircraft deviation from the designated flight level in cruise flight.

Additionally, in the automated ATC system of the Moscow consolidated control area, the following warnings are generated:

- on speed deviations;
- aircraft failure to reach the designated flight level due to incorrect vertical speed (flight level failure);
- on potential flight beyond the prescribed flight level.

In the automated aerodrome ATC system of the Khabarovsk ACC, the following reminder messages can be generated:

– on the necessity to initiate the control transfer/acceptance procedure;
– on the necessity to transmitting the ACT message to the adjacent control center;
– on the necessity to verify the correctness of setting the SSR code.

Additionally, in the automated ATC system of the Moscow consolidated control area, the following reminder messages are generated:

– on failure to fulfill the control acceptance/transfer conditions;
– on the necessity to switch to the manual flight coordination;
– on the planned maneuver;
– on the start-of-descent point
– reminder messages, specified by the air traffic controller (on reaching the designated waypoints).

The reminder message can be displayed in the automatic mode if it is relevant to the expected ATC event and can be monitored—for example, the aircraft control transfer when reaching a certain waypoint. The NCWs and the reminder messages are displayed in the zero line of the aircraft monitoring log in yellow digits on the screen background. Figure 2.28 provides the example of the MONA warnings, and Fig. 2.29 shows the MONA reminder messages [2].

Monitoring over the Lateral Correlation to the Current Flight Plan

The monitoring over the lateral correlation provides detection of the aircraft deviation from the planned or prescribed route, heading or ground track angle. If the working procedure requires the air traffic controller to enter all the amendments to the planned route and the relevant clearances (including the direct route, prescribed heading and ground track angle) into the system, the system assumes that any detected deviation from the may be regarded as the aircraft deviation from the air traffic controller's clearance (instruction), and consequently, generates the associated warning for the air traffic controller. In this case, the air traffic controller shall repeat the clearance (instruction) and enter the new aircraft movement into the system.

The MONA warns the air traffic controller on the aircraft deviations from the prescribed route, heading and ground track angle:

Fig. 2.28 Example of MONA warning display

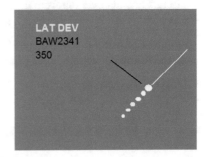

Fig. 2.29 Example MONA
reminder display

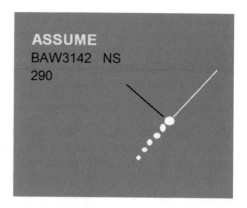

- for the sector of the aircraft flight control, the warning on the lateral deviation is generated if the lateral deviation of the aircraft flight from the prescribed route is beyond the pre-established allowable limits;
- for the aircraft with the prescribed heading and ground track angle, the warning generated for the sector of the aircraft control if the heading or the ground track angle in the track vector deviates from the prescribed heading and ground track angle for more than the pre-established limits allow;
- additionally, if the aircraft deviation from the prescribed route, heading and ground track angle is detected, the system shall (in EUROCONTROL procedures can be generated) generate a tactical trajectory of the deviation, based on the track vector of the track condition;
- in case the aircraft flies with the lateral deviation, when the air traffic controller enters a new prescribed route, heading and ground track angle, which correlate with the system track; the lateral deviation warning shall be removed;
- if for the aircraft with the previously revealed lateral deviation, no deviation from the prescribed route, heading and ground track angle is detected; the lateral deviation warning shall be removed.

Monitoring over the Longitudinal Correlation to the Current Flight Plan

The purpose of monitoring over the longitudinal correlation is to provide a more precise time estimation for the planned and tactical trajectories and to warn the air traffic controller in case considerable deviations from the original estimated time require the flight re-coordination. The system shall also warn the air traffic controller on the relative deviation of the aircraft speed from the prescribed. The system shall provide a more precise longitudinal estimation of the trajectory in accordance with the actual aircraft movement. The system shall monitor the longitudinal deviation for every correlated flight from its planned trajectory with the rate of at least 1 time per 60 s. Additionally, the system can check the longitudinal deviation of the aircraft flight position from its tactical trajectory for every update of the system track data. If the deviation of the longitudinal aircraft position, determined on the basis of the track vector from the expected longitudinal position, determined on the basis of the planned or tactical trajectory is above the allowable limit pre-prescribed for this class of trajectory, the system shall re-calculate the trajectory, using the track vector. The automatic re-calculation of the trajectory, without displaying the associated warning for the air traffic controller, shall be initiated if the detected longitudinal aircraft position deviates for more than 2 km (VSP) from the expected position.

Monitoring over the Vertical Correlation to the Current Flight Plan

The system monitors the aircraft movement in the vertical plane in order to warn the air traffic controller on a deviation from the prescribed flight level and to provide more precise calculation of the planned and tactical trajectories. If the aircraft deviates from the designated flight level, the system shall warn the air traffic controller accordingly (the prescribed flight level has not been reached). The system shall generate a warning for the sector of the aircraft flight control, if the actual flight level (AFL), shown by the track vector, verified at every update of the system track, is within the allowable limits from the cleared flight level (CFL) while the indicated vertical speed is above the allowable limit. The system shall generate the warning on the deviation from the flight level to the sector of the aircraft control, if the time period after the monitoring the aircraft on the CFL is above the prescribed allowable limit, with that the AFL in the track state vector deviates from the CFL for more than the prescribed allowable limit and the indicated vertical speed equals zero. The system shall generate the warning on the deviation from the CFL if the aircraft continues to perform the maneuver after reaching the prescribed flight level. If the air traffic controller has entered the new CFL for the aircraft, which has developed a deviation from the prescribed flight level or which has flown beyond the prescribed flight level, and the system track has become correlated, the associated warnings shall be removed. If the aircraft flight, for which the deviations from the CFL were present, no longer shows any deviations, the associated warnings shall be removed.

The Reminding Message for the Initiation of the Aircraft Control Acceptance/Transfer Procedure

The system shall generate a reminding message on the aircraft control acceptance/transfer (on the change of the control sector frequency) for the sector of the aircraft control at the prescribed time prior to the calculated time of reaching the coordination point if such a control acceptance/transfer procedure has not been initiated.

The reminding on the aircraft control acceptance/transfer procedure shall be removed when the frequency change has been entered or when the aircraft control has been accepted within the link instructions, or when the conditions for the generation of such a reminder no longer exist.

Reminding Message for the ACT Transfer to the Adjacent Control Unit

The system shall generate a reminding message for the transferring ATC unit on the ACT transfer to the adjacent control unit (by means of loudspeaker communication), which is not equipped with the instrumental OLDI interface with the system. The reminding message shall be displayed in the zero line of the aircraft monitoring log and in the control acceptance/transfer menu at a prescribed time prior to reaching the control transfer point (the suggested default value is 8 min, VSP). The reminding message is cleared from the screen at the designated time of 30 s (VSP) after the control transfer point has been reached.

Monitoring over the Correct Designation of the SSR Code

The system shall check the correctness of the SSR codes' designation and generate a reminding message in case the SSR code must be changed due to the following circumstances:

- the code is forbidden for input, the air traffic controller has entered the incorrect SSR code by mistake;
- no vacant codes are available at the moment;
- the two identical codes are present in the system;
- if the ACT message contains the SSR code, which cannot be used in this control area, the system suggests a vacant code, which is displayed for the air traffic controller in the zero line of the aircraft monitoring log for its further transmission to the aircraft;
- in case the system receives Code 2000 from the aircraft (i.e., no code has been designated).

2.3.3 Medium-Term Conflicts Detection (MTCD)

General Information on MTCD The MTCD is an aid for planning non-conflict trajectories for the air traffic controller of the radar control unit (RCU) and the air traffic controller of the procedural control unit (PCU) with the common time of detection within the range of 0–20 min. The areas of responsibility of the RCU and PCU are divided. The RCU resolves the conflicts if there remain no more than 8 min prior to the conflicts occurrence, when the aircraft is within their control sector. The PCU solves the similar tasks when the aircraft enters their control sector or when the aircraft has not entered the control sector of their responsibility within the prediction time period of 8–20 min (VSP). The purpose of the MTCD is to inform the air traffic controller on the conflicts which occur in the middle-term period and which may require to amend the aircraft flight plan or to issue another clearance. The warning on the conflict within the range of 0–2 min is provided by the STCA function. The MTCD includes the following three functions:

(a) detecting and informing of the air traffic controller on the possible loss of the prescribed separation between the two aircraft. With that it is also important to take into account that the conflict is a probable phenomenon, an accidental event. The ambiguousness of the conflict occurrence increases with the increase of the time period for which the prediction is made;
(b) detecting and informing the air traffic controller on the aircraft which is intersecting into the isolated or by any other ways restricted airspace;

(c) detecting and displaying the closing-in movement of the two aircraft for the air
traffic controller, when each of the aircraft is blocking the airspace which may
be used by the other aircraft, though the prescribed flight level separation will
be reached, for example, in case of the pilot's request on the alternate flight level
or in case of the conflict resolution, affecting one of the aircraft. In other words,
the MTCD ensures the air traffic controller is informed on the airspace areas,
where aircraft block other aircraft's movement.

The MTCD function detects:

– aircraft conflicts: the shrinkage of the separation between the probable positions of
the two aircraft, based on the system trajectories, taking into account the ambiguousness of the prediction;
– context conflicts;
– transgression into the special-use airspace: non-compliance with the distance
requirements between the probable aircraft positions and the limited-use airspace.
The detection range is 0–20 min;
– descent below the lowest allowable flight level: probable aircraft positions in the
airspace below the lowest usable flight level within this airspace. The detection
range is 0–20 min.

MTCD Display For the display of the MTCD, a set of aids, providing the graphical representation of the actual and forecast air traffic situation, potential problems
and airspace problems, are used. These aids, both for horizontal and vertical representation, based on the data from the current flight plan, updated by the air traffic
controllers and/or by the system, include:

1. potential problem display (PPD).
2. vertical assistance window (VAW).
3. horizontal projection of the trajectory, shown in the air traffic situation window.
4. symptom of the detected problem in the aircraft monitoring log (●) and in the
"PLANNING LIST" of the MTCD lists.

The PPD and VAW are called out on the ASD by means of the associated keys of
the icon window.

Potential Problem Display (PPD). Purpose and characteristics of the PPD

The PPD is an aid for graphical and list representation of the potential problems and
the airspace problems, related to the given sector and the certain buffer area around
the sector (VSP) within the range of the successive 20 min (VSP). The example of
the PPD is provided in Fig. 2.30.

The PPD is dynamically updated every time when the information is updated or
due to the air traffic controller's inputs.

The PPD contains the following elements:

– list of the potential problems representation;
– graphical representation of potential problems;

Fig. 2.30 PPD window

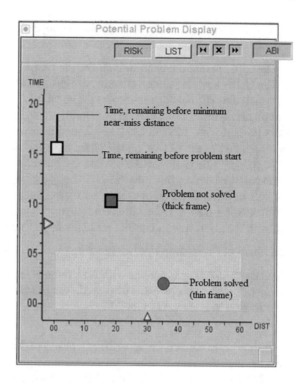

– elements for the display control.

  The PPD has the following characteristic features:

– non-transparent, with the frame and the title;
– can be shifted;
– can be modified in size with the control element which is located in the right lower corner (press any mouse pushbutton and, holding it pressed, shift it upward, downward, to the right or to the left);
– can be iconized if it contains no information. If a problem occurs, the display expands automatically;
– on default when the control panel is switched on:
– the widow is opened on the ASD;
– the detected conflicts and risks of conflicts in the control sector as well as the problems which occur 2 min prior to the aircraft entry into the sector and within the 2 min after the aircraft leaves the sector;
– the window is provided in the graphical format.

  The PPD displays conflicts and risks of conflicts. The problems are classified as follows:

– as "conflict" if the problem occurs between the actual flight level (AFL) and the cleared flight level (CFL) (or on the entry flight level if the aircraft has not entered the sector);

– as "risk of conflict" if the problem occurs beyond the CFL (in the altitude layer from the CFL to the transition flight level (TFL).

The upper part of the window contains the "RISK," "LIST/GRAF" keys, ⊬ ⊠ ⊯ .
The "ABI" key ("Advanced Boundary Information") is used to display the flights, for which the ABI message is received. The "RISK" key is used to recall/clear the information in the PPD window on the aircraft, for which the risk of conflicts is graphically present. The "LIST/GRAF" key is used to switch the form of the information representation in the PPD window. When the window displays the information as a list, the key title runs "GRAF"; when the window displays the information in a graphical form, the key title runs "LIST." The ⊬ ⊠ ⊯ keys are used to clear the conflict symbols in the in the PPD window as follows: ◀for aircraft on reciprocal tracks, ✖for crossing tracks, ▶▶for tail-chasing (same) tracks. If the key is displayed as released, the given type of conflict is not displayed in the PPD window.

List Form of Presenting the Information

When the "LIST" key is pressed with the left mouse key (LMK) in the PPD window, the list of potential problems is displayed. The example of the list representation of the information is shown in Fig. 2.31.

The window contains the following data:

– the time scale tailored for up to 20 min (VSP) of the information display;
– potential problem symbol;
– problem log, if recalled (the description id provided below);
– amber triangle on the time axis for limiting the airspace area, if the problem occurs in this area, the associated point is automatically recorded in the aircraft monitoring log.

Graphical Form of the Information Representation

When the "GRAF" key is pressed with the left mouse key (LMK) in the PPD window, the potential problems and the airspace problems are displayed in the graphical form. The example of the graphical representation of the information is shown in Fig. 2.32.

The area of responsibility of the RCU air traffic controller (from 0 to 8 min) is highlighted in light gray.

In the graphical form of the information representation in the PPD window, the coordinate axes are used for the following purposes:

The X-axis shows the minimum separation distance, presenting the distance in a kilometer range of 0 up to 100 km (VSP). The scale has a grade of 5 km.

– The range of the minimum distance presentation can be changed as follows: Press the LMK on the X-axis and, holding it pressed, change the range by moving the mouse to the right or to the left. The zoom-out or zoom-in of the displayed minimum range expands or shrinks the area of the displayed potential problems;
– The Y-axis shows the time of reaching the designated waypoints, displayed in minute intervals of 0–20 min (VSP) with the marking of each minute. The time scale can be changed similarly as described for the X-axis.

Fig. 2.31 List representation
in PPD window

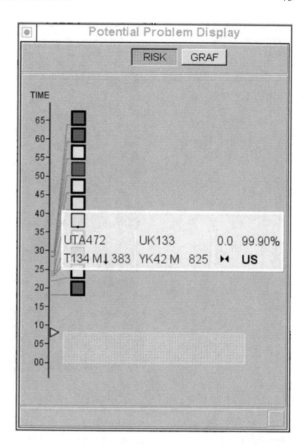

Fig. 2.31 List representation in PPD window

The given coordinates show the symbols of potential problems. The symbol and its characteristics are similar to the symbols in the list-form representation of the information. The positions of the symbols are updated every 30 s (VSP) or by the air traffic controller's inputs.

In the graphical form, the symbols shift downward in the direction of the decrease of time, remaining for reaching the minimum separation distance between the two aircraft. If the flight plans of the conflicting aircraft are not amended, the symbols move vertically downward with every updating. If a change, affecting the minimum separation distance between the two aircraft, is detected, the symbols also shift horizontally. The problem symbol can be displayed in the form of a square, a diamond, or a circle. The potential problem is displayed with a square or a diamond with the black frames in the X-axis coordinate, related to the predicted minimum separation distance between the two conflicting aircraft in kilometers, and in the Y-axis coordinate, related to time in minutes, remaining before the conflict (the lower frame). The square is changed for a circle, when the potential problem enters the tactical airspace of the RCU and the time remaining before the problem is less than 8 min (VSP).

Fig. 2.32 Graphical form of information display in PPD window

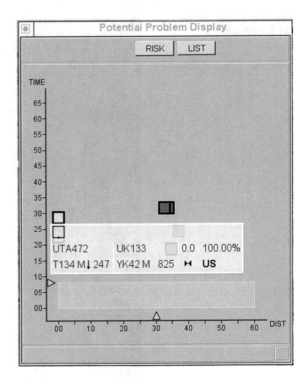

The airspace problem is displayed as a diamond in the X-axis coordinate, related to the predicted minimum separation distance in kilometers, and in the Y-axis coordinate, related to time, remaining before reaching this point (in minutes). The airspace problems shift subsequently in the PPD window with every updating.

The risk of conflict is displayed in a brown frame.

The amber triangles on the time and distance axes are used to limit the airspace area where the entered problem will automatically trigger the problem character (the red dot) in the aircraft monitoring log. The position of the triangles on the axes is modified by the air traffic controller. To perform it, press the right mouse key (RMC) and shift the mouse, holding it at the triangle. With that, the dynamically changing value of the changed parameter is displayed below the triangle. If the problem symbol is positioned below the amber triangle on the time axis and to the left of the amber triangle on the minimum separation axis, the red dot is displayed in the aircraft monitoring log, which stands for a medium-term conflict and attracts the air traffic controller's attention. The amber triangle can be set for any value.

If the minimum separation distance is less than the standard separation, a line, parallel to the time axis, is displayed above every potential problem. The line starts at the upper part of the potential problem symbol, as shown in Fig. 2.33, and ends in the point, associated on the Y-axis with the time, remaining before reaching the minimum separation distance.

Fig. 2.33 Potential problem
symbols in PPD window

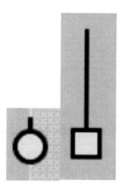

When the mouse curser is pointed in the problem symbol, the color of aircraft monitoring log symbols of the conflicting aircraft changes on the ASD and in the Zoom window. When the LMK is pressed on the problem symbol, the trajectories of the conflicting aircraft are displayed on the ASD. The symbol color and frame shape indicate as follows:

– The problem has been either observed or not (thick or thin frame respectfully);
– If the problem has a high probability of occurrence (more than 0.95, i.e., more than 95%), the problem is shown in yellow ☐;
– If the problem has a low probability of occurrence (less than 0.0001, i.e., less than 0.01%), the problem is shown in gray ☐;
– In other cases (the probability of occurrence is from 0.01% to 95%), the problem is shown in blue ☐;
– If the problem has an STCA status, it is shown in red ●.

For the air traffic controller, it means the following:

– the yellow color: A decision on the interference into the aircraft movement shall be made in order to resolve the conflict (the conflict occurrence probability is high);
– the gray color: Less attention may be spent on the potential conflict (there is a high probability that no conflict will occur);
– blue color: wait, as the conflict can resolve by itself (additional attention is required; there is an ambiguity of the conflict occurrence).

The problem is displayed as observed, when the air traffic controller observes it and presses the LMK or the RMK on the symbol of this potential problem in the PPD or VAW window.

For example,

– the problem has not been observed: thick black (or brown—for the risk of conflict) frame ☐;
– the problem has been observed: thin frame ☐.

In case the STCA annunciation has been switched on for the potential problem, the symbol shall be displayed in the field of the ALERT color. This displaying shall have a priority over any other indications.

Fig. 2.34 Problem log

Description and the Control Panel Operations with the Problem Log

When the mouse curser is shifted on the symbol of the problem or the airspace problem in the PPD window, a framed box temporarily opens near the problem symbol—a problem log with the text lines of the associated information.

An example of the problem log is provided in Fig. 2.34.

Simultaneously, the color of the aircraft monitoring log symbols on the ADS is changed for red.

The problem log contains as follows:

The first line: the aircraft call signs (flight numbers); predicted minimum separation distance between the two aircraft in the intersection point (in kilometers); the probability of the problem occurrence (in percentage).

The second line for each aircraft: the aircraft type; the aircraft weight category in accordance with the vortex wake turbulence; trend of altitude change; prescribed flight level; type of conflict in the form of a symbol (same (tail-chasing) tracks, crossing tracks, reciprocal tracks, etc.); the symbol for the point of conflict, if available.

The conflict symbol can be as follows:

▶▶ same (tail-chasing) track;

◀ reciprocal tracks;

✕ crossing tracks;

➤ merging tracks (not mandatory to be displayed).

If the conflict (risk of conflict) is related to the problem between the aircraft and the limited-use airspace area, the second log in the problem area contains:

- the indication of the limited-use airspace area (in the first line);
- the lower and upper flight levels of the limited-use airspace area (in the second line).

When the LMK is pressed on the problem symbol in the PPD window, or in the problem log or on the conflict symbol in the aircraft monitoring log on the ASD, the routes of the conflicting aircraft are displayed in green with the indicated route turning waypoints, as it happens when the planned route is recalled, with the colored areas of the route from the start to the end of the separation transgress. If the conflicting aircraft are chosen, these areas are colored in red. If the air traffic controller chooses the aircraft, for which the risk of conflict is detected, the color is amber. When the LMK is pressed on the problem log, the log is displayed in the continuous mode. When the LMK is pressed on the log for the second time, it is removed from the screen. When the mouse cursor is placed on the field with the aircraft call sign, the aircraft monitoring log for this aircraft is illuminated in the problem log on the ASD. Simultaneously, the aircraft monitoring logs for the involved aircraft are illuminated.

Vertical Assistance Window (VAW)

The vertical assistance window (VAW) is a graphical aid, showing the vertical profile of the planned trajectory of the selected aircraft via the sector, starting with its actual position up to 2 min (VSP) after the aircraft leaves the sector. In case a potential problem, related to the selected aircraft, is detected, the VAW also shows other aircraft, conflicting with the selected aircraft. This window assists air traffic controllers in their analysis of the predicted air traffic situation. The selected aircraft (in the VAW context) is an aircraft, selected for the analysis in the VAW. The selected aircraft's call sign is displayed in the upper part of the VAW.

The conflicting aircraft are the aircraft, which, according to the prediction, will be conflicting with the selected aircraft. The context (limiting) aircraft are the aircraft, which provide the limitations for the selected aircraft's maneuvering. The context aircraft may be hazardous in case the trajectory of the selected aircraft is changed. The area of the context aircraft of the area of a potential conflict, the area of the airspace which is blocked by the context aircraft for the movement during a certain predicted moment of time. The VAW displays the air traffic situation in regard to the selected aircraft. The example of the VAW is shown in Fig. 2.35.

Fig. 2.35 VAW

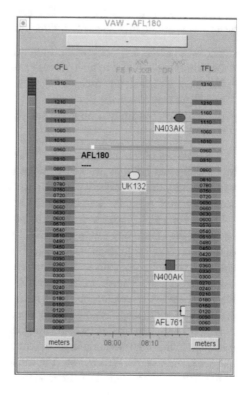

The VAW has the following characteristic features: non-transparent, with a frame and a title; shiftable, its size can be changed by means of control elements which are positioned in the right lower corner (press any mouse key and, holding it pressed, shift it upward, downward, to the right or to the left); it can be recalled from the icon window and iconized by pressing the associated icon key with the LMK.

When the VAW opens, it is automatically updated by the information received from the MTCD, or as a result of the air traffic controllers' inputs (in case of their reference to problems). When the VAW is closed, it shall come out automatically onto the ASD in case the problem occurs to the climbing or descending aircraft as well as when the mouse cursor is pointed onto the problem symbol in the PPD window. The VAW can be called out not depending on whether a problem exists or not. The VAW shows context aircraft (vertical profile).

The VAW contains the following elements:

- the title with the name of the window and the call sign of the aircraft, for which the VAW has been called out;
- the aircraft call sign box for selecting the aircraft;
- the VAW upper part elements, linked to the time scale by thin vertical lines;
- the Y-axis: two altitude axes with the shown flight levels on each side of the window, the flight levels on the left side stand for the CFLs, the flight levels on the right side are the TFLs. The aircraft flight levels are shown as thin single horizontal lines.

The currently selected CFL and TFL are highlighted. The eastward and westward flight levels are highlighted with lighter and darker shades, respectively;

- the "METERS" keys below the time scale for switching the flight level values from hundreds of feet to tens of meters and backward;
- the selected aircraft position symbol with the aircraft monitoring log, speed vector, and flight trajectory;
- the areas of the limited airspace for the aircraft, conflicting with the selected aircraft.

The call-out of the information on the selected aircraft in the VAW is performed by pressing and holding the RMK on the field of the aircraft call sign in the PPD window or by the aircraft call sign field in the upper part of the VAW. The air traffic controller shall press the key below the window title with the LMK, the key will split in two: The left part will show the pop-down sliding menu, which contains a list of all the conflicting aircraft; the right part will display a box for manual entry of the aircraft call sign by means of a keyboard. The double pressing on the selected call sign from the list with an LMK will lead to the display of the information on associated aircraft. The air traffic controller can enter a new CFL or TFL by pressing the LMK on the selected flight level field. With that, the background color of the pressed key will change from gray to green, and the number digits will be displayed in white. There is an option of controlling the X-axis scale for scale changing within the range of the displayed minimum and maximum possible values. For this purpose, the LMK must be pressed and shift to the left or to the right when held. To the left of the CFL column, there is a slider with two moving indexes, which stand for the displayed current upper and lower flight levels for changing the flight level scale within the range from minimum to maximum flight levels. The selected limits are shown on the left side of the respective moving index. The upper and the lower limiting flight levels of the airspace of the control sector are displayed at the ends of the sliders. The area between the two moving indexes stands for the displayed current flight level and is displayed in a darker shade than the slider. Moving upward and downward along this area allows viewing all the flight levels within the limits of the airspace area with the fixed flight level scale. The selected aircraft position, indicated in the vertical profile of the flight by means of the associated symbol, is dynamically updated. Near the symbol, the aircraft monitoring log, containing the aircraft call sign and the current flight level, is displayed. The speed vector is displayed in accordance with the currently selected parameters in the display management window. The flight trajectory of the selected aircraft is displayed in the VAW by means of a green line throughout the entire sector; its starting end is marked by the symbol of the current aircraft position.

The vertical profile of the aircraft flight is marked with the intersection points of the selected aircraft with the conflicting aircraft in form of areas of the limitations of the airspace, marked with the call signs and highlighted in color. The color of the limitations of the airspace depends on the conflict probability and correlates with the colors of the symbols in the PPD window.

Conflict-Oriented Horizontal Trajectory Projection

The call-out of the conflict-oriented horizontal trajectory projection is used for viewing the air traffic situation in case of a problem in order to make a decision on the conflict resolution. The horizontal projection is displayed in the radar window and provides a graphical representation of the selected aircraft's trajectory throughout the entire route of its flight (throughout and beyond the sector). The horizontal trajectory projection is called out on the ASD by pressing the LMK on the problem symbol in the PPD window or in the VAW, or by pressing the LMK on the red dot in the aircraft monitoring log. The display of the conflict-oriented horizontal trajectory projection can be provided in the time-view mode (by pressing and holding the LMK on the problem symbol in the PPD window or in the VAW). Conflict-oriented horizontal trajectory projections contain the flight routes of the conflicting aircraft starting from the current aircraft positions and throughout the entire length of the sector. The trajectories for the both conflicting aircraft are displayed in green continuous (unbroken) lines starting from the current aircraft positions and running to the of the maximum closing point with the indicated route turning waypoints as well as with the information on time and flight levels of their passage. In case the minimum distance between the aircraft is below the standard separation minima, the parts of the trajectories, running from the start of conflict to the maximum closing point, are displayed in red. In case the minimum distance between the aircraft is above the standard separation minima, no red trajectory parts are present. The example of the conflict-oriented horizontal trajectory projection is provided in Fig. 2.36.

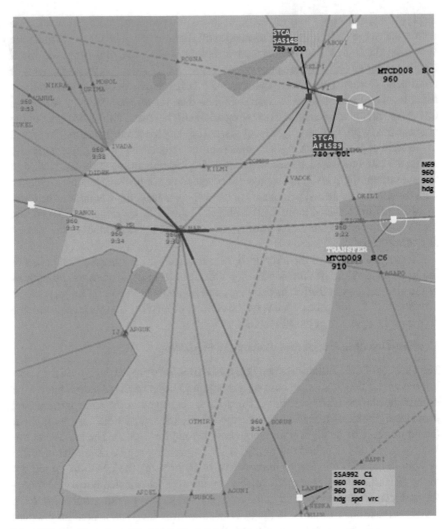

Fig. 2.36 Conflict-oriented horizontal trajectory projection

2.3.4 Automated Coordination and Control Transfer

General As long as the aircraft flight proceeds, the ATS units transmit the necessary information on the aircraft flight plan from one unit to the other well in advance, in order to ensure that the accepting ATS unit will be able to obtain the data, analyze them, and perform the necessary actions coordination. In order to minimize the coordination by means of voice communication, the ATS units establish and use standard rules of coordination and flight management, which are enshrined in the Letter of Agreement (LoA) or in the local instructions (within the FIR). The responsibility for the aircraft control is usually transferred when the aircraft crosses the control area boundary, but the can also be transferred in any other point close to the boundary or at any other time prior to the boundary crossing, which shall be agreed in-between the two ATS units. Note: There is a difference between the terms "communication transfer" and "control transfer." The latter case, in spite of the established communication between the aircraft and the accepting air traffic controller, the responsibility for the control is not taken over until the aircraft crosses the control area boundary or any other point of control transfer, agreed in the LoA between the two ATS units. Further, for the sake of facilitation, the term "control transfer" will be used if there is no intension to mention specifically the communication transfer. Depending on the circumstances, such LoAs and instructions can include, among the others, the following elements [3]:

– designation of the areas of responsibility and mutual interests, of the airspace structure and classification;
– the rules for the flight plans and air traffic control data exchange, including the use of coordination messages, transmitted by means of automated and/or voice communication aids;
– communication aids;
– main points, flight levels, and time of control transfer;
– main points, flight levels, and time of communication transfer;
– control transfer and acceptance conditions, such as established altitudes/flight levels, the exact values of minima or separation intervals for the control transfer, the use of automated aids;
– the rules which shall be applied in case of unforeseen conditions.

The process of coordination and control transfer officially consists of the following stages:

– notification on the aircraft flight for the purpose of ensuring the readiness of the coordination if necessary;
– the coordination of the control transfer from the transferring ATS unit;
– the coordination (if necessary) and the acceptance of the control transfer conditions by the accepting ATS unit;
– the control transfer to the accepting ATS unit;
– control acceptance.

The conventional process of coordination with the application of the radio communication in-between the two ATS units includes a notification, communication, and reaching the agreement. In standard conditions, it can be confined to a simple notification on the entry conditions with the simple acknowledgment, while in other cases the agreement is reached by means of negotiating. General provisions on the coordination and control transfer are established by the ICAO and are provided in ICAO Doc 4444 (PANS-ATM), Chap. 8, 8.7.4, Chap. 10, 10.1.2, and Chap. 11, 11.3.7, *Air Traffic Management: Procedures for Air Navigation Services* [3].

The ICAO Doc 4444 [3] prescribes the standard phraseologies and types of coordination for use in different situations. During the voice coordination by means of the loudspeaker communication, several coordination elements are usually embraced; with that, there may be no precise segregation of the separate elements, which belong to different stages of coordination, mentioned above. Within the FIR of one ATM unit, the control transfer between the two adjacent control sectors may be provided (and usually is provided) without the preliminary coordination, if all the conditions below are fulfilled:

- standard conditions of control transfer, i.e., the compliance with local instructions;
- the aircraft positions are displayed on the ASD in accordance with the data from the SSR (and/or form the ADS-B) with the provided aircraft monitoring logs;
- prior to the control transfer, the aircraft is identified on the ASD by the accepting air traffic controller;
- the air traffic controllers are provided with the constant two-way direct voice communication aids, which allow them to literally immediately establish a communication link with each other.

SYSCO Concept and Levels The process of coordination and control transfer between the two ATS units is significantly facilitated if there is an established set of standard messages, for the transmission of which the digital data transmission links are used [4, 5]. In the year of 1997, the EUROCONTROL have developed a concept of the automated message exchange, which was entitled "system-supported coordination," or "SYSCO."

The main principle of the SYSCO concept is to establish the so-called area of common interest, or "ACI," in which for the both ATM units, the information on all the air traffic from both sides of the common control area boundary of the FIR is available, similarly to the process which is run within FIR, i.e., at both sides of the control area boundary the information on the aircraft position with the aircraft monitoring logs, including the ATFM plan information elements, is available. The type of the air aircraft monitoring log indicates the status of the aircraft control and management. Simultaneously, the aircraft trajectories are calculated, viewed, and controlled for the consecutive n-number of minutes (up to 20 min) of flight (the TP and MONA functions), and the MTCD function can detect the aircraft conflicts within the boundaries of another ATM unit sector much prior to the time, when the aircraft enter this sector. This allows the preliminary coordination of such conditions of the aircraft leaving of the control area and its control transfer to the next sector that these conflicts could be avoided. These tasks are solved in the modern automated ATC systems and are usually laid on the planning controller (PC),[1] working together with the tactical controller (TC).[2] Thus, the SYSCO determines the operational concept, based on the improved automated ATS aids, which provide as follows:

---

[1] In the automated ATC systems of the Russian Federation, the similar function is laid on the air traffic controller of the procedural control unit (PCU).

[2] In the automated ATC systems of the Russian Federation, the position corresponds to the position of the air traffic controller of the radar control unit (RCU).

- processing of the observation information (radar and ADS data);
- processing of flight (FM plan) information;
- displaying the air traffic situation at high resolution and the color-based coding of the flight control status;
- processing the data on presence or absence of the standard control transfer conditions;
- improved data storing;
- precise prediction of an aircraft trajectory for 20 min in advance;
- high-speed data exchange between the ATS units for the purpose of automated exchange of standard messages in form of a dialog.

More precise messages, based on the actual system information on the flight trajectory, specified by the observation data, ensure the air traffic controllers obtain more reliable information on conditions of the aircraft entry into the controlled airspace, which leads to the decrease of the workload when performing the coordination and control transfer tasks. At the same time, the improved accuracy and integrity of the data allows the application of the RVSM while ensuring the same level of flight safety. The SYSCO concept implies the gradual transition from the initial-stage functional capabilities to their full-scale implementation. Every stage of implementation operates with a certain set of functional capabilities, which is gradually improved, while the main coordination and control transfer events remain the same throughout every stage and on all levels. In fact, every level of SYSCO implementation is determined by the minimum capabilities of the coordination partner's system. This approach is clarified in Fig. 2.37. The basic level of the Pre-SYSCO provides the basic automation of the coordination process, thus considerably decreasing the workload of the air traffic controller and providing the option of transmitting the coordination information in form of flight plan data and aircraft control transfer conditions by means of the electronic equipment. All the data transmissions are confirmed upon reception by the corresponding ATS unit, where these data shall be applied, and as for the data, they are generated and transmitted automatically on the basis of the pre-defined parameters, agreed by the two sides.

The Pre-SYSCO is a method of transmitting the data "in one direction" (i.e., there are no aids for supporting the dialog exchange); it has a limited number of aids for system-supported air traffic controller interface and does not provide full amount of information for the air traffic controller and the support system. For the Pre-SYSCO level implementation, the data input aids, the aids for viewing the received data and the associated warnings, the basic system of flight plan processing and automatic generation of data for transmitting as well as the aids for the message exchange are required. The SYSCO 1 level provides the intermediate stage of automated coordination and aircraft transfer aids, its aim purpose is to implement the significant operational and technical aspects of the full-scale SYSCO concept, thus ensuring further improvement of the airspace use and decreasing the air traffic controller's workload.

SYSCO 1 is more complex than the Pre-SYSCO level concerning the provision of far more options for the air traffic controller's interface, supported by the system,

The aim is decreasing the workload and increasing the efficiency of the airspace use:	SYSCO implementation stages			
	Basic level (Pre-SYSCO)	SYSCO 1	SYSCO 2	SYSCO 3
The aircraft is part of the automated coordination with the use of the "AIR-GROUND" digital data links (DDL) Controllers are provided with additional information and decision-making support at different stages of the coordination and control transfer (CORA).				▓
Almost complete automation of ground coordination and control transfer: improved observation and flight plan processing; implementation of the area of common interest (ACI); improved flight trajectory prediction (TP); improved conflict detection (MONA).			▓	▓
Intermediate stage of automated coordination and aircraft control transfer: further decrease of verbal coordination; control transfer functions; basic flight trajectory prediction (TP); detection (filtering) of compliance/non-compliance with the Agreement conditions; basic observation.		▓	▓	▓
Basic automation of the coordination in form of automatic flight plan data and aircraft control transfer conditions transmission with the confirmation of receiving (ABI, PAC, ACT, REV, MAC, LAM messages; see below).	▓	▓	▓	▓

Fig. 2.37 SYSCO implementation stages—automated coordination and aircraft control transfer

thus further reducing the verbal coordination and providing a technical basis for a more complete automated support; however, this level still does not provide a full-scale information for the air traffic controller's support. To reach the SYSCO 1 level, the ATS units shall implement all the minimum functional capabilities of the Pre-SYSCO level + basic observation for the purpose of providing the systems with the information on aircraft position, dialog-supporting aids, for the purpose of decreasing the number of the remaining events of verbal coordination and control transfer, basic trajectory prediction and control transfer functions.

The SYSCO 2 level provides complete SYSCO automation, contributing the utmost for the improvement of the airspace use and for the decrease of the air traffic controller's workload. This level is more complex than SYSCO 1, as it provides the full-scale capability of air traffic controller's system-supported interface, thus decreasing the verbal coordination to its minimum. This level also provides full information for the air traffic controllers' and systems' support, significantly decreasing the workload and providing more efficient use of the available airspace. To reach the SYSCO 2 level, the air traffic controllers shall fully implement the functional capabilities of SYSCO 1 as well as:

– display of the area of common interest (ACI) and full-scale implementation of the option of data processing in the flight operations conditions;
– improved observation and flight plan data processing as well as functions of trajectory prediction and conflict detection;
– system-initiated function of aircraft control transfer.

SYSCO 3 provides still more improved functionality. This level implies further improvement in airspace use and in decreasing the air traffic controller's workload. The SYSCO 3 level implies the development of such technical aids as aids for providing the air traffic controllers with additional information and support in decision-making process at different stages of the coordination and control transfer (CORA), the "ground–air" DDL in order to involve the final partner of the coordination, i.e., the pilot, into the automated coordination process, thus decreasing the common coordination workload. The "ground–air" DDL can also improve the preciseness of the keeping the common database which will contribute to the further improvement of the airspace use.

OLDI Standard of EUROCONTROL

The all-Europe implementation of the multi-level SYSCO concept is realized by the OLDI[3] Standard, which was developed in 1992 by EUROCONTROL for establishing the principles of message exchange among the ATS units in order to provide automated coordination and control transfer from one ATS unit to the next [6]. In order to reach this aim, the OLDI establishes standard procedures when the movement of each aircraft through the control area boundaries is coordinated in advance, while the aircraft control transfer occurs at crossing the above-mentioned boundary or in its vicinity. The EU states started to apply to this Standard in the early 1990s, mainly among the ATM units. Basically, the OLDI Standard ensures the merging of the flight data processing systems (FDPSs) of different area control units. In order to contribute to the SYSCO implementation [5], common rules and formats of messages were developed. Later on, the OLDI Standard has been constantly developing, embracing higher and higher levels of ATS units' dialog automation. The currently used version is SPEC 4.2 (of 2012), which includes:

---

[3]On-Line Data Interchange.

- basic type of messages:

  - basic procedure messages [ABI, ACT, REV, MAC, PAC, and logical acknowl-edgment messages (LAM)];
  - messages on the air traffic situation in the vicinity of the boundaries of the given ATS control areas (two types of messages);
  - messages on coordination among the ground-based civil aviation and military aviation units related to the organization of passing the associated airspace (seven types of messages);
  - messages for communication transfer procedure (two types of messages);
  - messages for maintaining the established communication with the aircraft via the DDL (two types of messages);

- additional messages:

  - for the basic procedure;
  - for the communication transfer procedure;
  - pre-flight coordination messages;
  - dialog coordination messages;
  - information messages for third parties;
  - messages for the coordination among in-between the oceanic ATS units and area control units.

The OLDI Version 4.2 provides the total of 31 message types (not including the replies), but it does not mean that all of them are currently used. In 2011, the EUROCONTROL issued the document, entitled "SYSCO Concept Implementation Manual," which provides a list of 21 types of messages, which are included into the European Commission's EC Regulation No. 1032/2006 on the implementation of the OLDI Standard came into force on December 31, 2012. The EUROCONTROL's SESAR program, the ICAO's Global Operational ATM Concept (ICAO Doc 9854) [9], implies an even wider implementation. The information exchange is based on the presence of associated and correlated communication protocols, which provide their reliable interoperability. For message exchange, the EUROCONTROL suggest to use the ATS Data Exchange Presentation (ADEXP) format, which establishes the formatting principles, grammar rules, etc. The abbreviated format, defined in the ICAO Doc 4444 [3], is also allowed for use. At present, the OLDI Standard is widely used for notification and initial coordination of all the ACCs of the European Union (EU) for communication with the adjacent ACCs, as well as between the ACCs and the aerodrome control units. The OLDI implementation is maintained by the application of the control acceptance/transfer messaging and the dialog procedures for tactical amendments implementation (speed, heading, flight level, vertical speed). In the ATM of the Russian Federation, the basic level of the OLDI Standard is applied for the communication among the consolidated ATM centers, and the EU, Ukraine, and Belarus ATM centers.

Initial Coordination and Basic Level of OLDI

The exchange of the estimated data by means of the OLDI, while all the other coordination is performed in voice, is known as the "basic OLDI level." According to ICAO Doc 4444, the successive ATS unit, according to the flight route, shall be provided with a notification on the flight, and as a result of the coordination process, an operational agreement on the transfer conditions shall be obtained. In order to decrease the demand for coordination, ICAO Doc 4444 recommends the ATS units to establish LoAs and to apply the standard procedures on coordination and control transfer. Typically, the standard routes and flight levels are applied. In such situations, the initial coordination for every flight with respect to such conditions will confine to simple transmission of the estimated data (or "estimates") and the confirmation upon their receipt. In case of using the telephone communication, the initial coordination is provided approximately 30 min prior to the estimated time of the aircraft boundary crossing of the FIR boundary. First the notification message (ABI) is transmitted, followed by the more precise information in the activation message (ACT). These messages are transmitted automatically at a time, specified in advance by the LoA, prior to the boundary crossing of the FIR. As for the ACT message, it is normally 10–15 min prior. The air traffic controllers of the ATS units, which are higher in position, normally have an opportunity to initiate the data transmission even earlier, if they have an intension to do so, and if they do not envisage any evidence for amending the transfer conditions (route and flight level). In case some problems occur concerning the transfer, the basic OLDI Standard shall be supported by the loudspeaker communication, when, for example, no LAM confirmation is received. The voice coordination is also applied, when the air traffic controllers of the lower sector are not able to assume the aircraft control on the suggested transfer conditions, or every time when they need to apply the non-standard conditions for the aircraft control transfer. Below, in the section, entitled "Basic types of messages for the basic procedure," the information on the basic procedure messages, i.e., messages which do not require and coordination, is provided.

Control Transfer Functions

Conventionally, and as described in the ICAO Doc 4444 [3], after the initial coordination, the aircraft control transfer implies the transmission of the aircraft (radar) identification index, the communication transfer, and the transfer of control. With the use of the modern observation systems, the aircraft identification index rarely requires coordination. The communication/control transfer also rarely requires coordination if the airspace is well-organized and the mutual agreements are obtained. With the SYSCO implementation, there are functions, indicating the air traffic controllers not only for the sectors of their control, but also for the other control sectors of other ATS units, when the aircraft flight crew receive an instruction for contacting the consequent air traffic controller at the associated frequency as well as when this contact has been established. Other messages provide the detailed information on the updates, resulting from the actions of control, which, probably, may affect the control transfer process. There are also messages for earlier control transferring, or for the aircraft control transfer on the suggested conditions. Many of the present ATS systems use the simple "ASSUME" input for the purpose of highlighting the aircraft, accepted in "their" sector, i.e., the aircraft which have switched to the frequency of the control sector of their responsibility and are under this sector's control. In the automated ATC systems of the Russian Federation, the automated procedures of communication/control transfer among the control units with the use of the OLDI messages are not applied, i.e., the associated message exchange is performed via the voice communication system. The system monitors the flight performance, and at the points of control transfer/acceptance, generates the associated reminder messages for the air traffic controller. As a result, the air traffic controller, by system inputs, confirms the agreed and performed actions on control transfer/acceptance, which affects the information in the aircraft monitoring log and in the list of aircraft flights.

Other Functions

The OLDI specification also includes the messages for coordination of the amendments to the individual elements of the flight plan, such as flight level change, nonstandard flight level or route, direct route, speed or vertical speed limitations, the civil/military interceptions of airspace of the other ministry, the requirements for holding or maintaining the speed in order to ensure the right arrival sequence, etc. Some messages provide certain flexibility in their application, and there are several alternative messages for coordination of certain conditions. The SYSCO implementation requires the two-party coordination of the applied functions and messages. With the adequate choice of messages and methods of their applications, it is possible to automate almost every aspect of coordination.

Note: In the automated ATS systems of the Russian Federation, the functions of the automated message exchange among the adjacent ATS units have not been implemented so far.

Main Types of Messages for the Basic Procedure

Table 2.1 provides the abbreviations of the message types, mentioned above, as well as the sequence of their application.

All these messages deal with the amendments in the active flight plan. Table 2.2 presents the flight plan field numbers, which are affected by the above-mentioned messages.

Logical Acknowledgment Message (LAM)

A LAM is a message, used by the receiving control unit confirms to the transmitting control unit, that the transmitted message has been received, checked to contain no errors, stored, and can be displayed at the air traffic controller's workstation. The LAM is automatically generated and transmitted immediately with no inputs from the air traffic controller, not depending on whether the flight plan, relevant to this message, has been found or not. If the transmitting control unit does not receive the LAM within a certain time period (timeout), a "**NO LAM**" warning is displayed

Table 2.1 Message decryption

Message type	Abbreviation	OLDI message specification
Advance boundary information message	ABI	Used for preliminary notification on the conditions of boundary crossing
Activation message	ACT	Used for activation in standard conditions
Logical acknowledgment message	LAM	Used for logical acknowledgment of received message
Preliminary activation message	PAC	Used for preliminary activation in standard conditions
Revision message	REV	Used for specification of transmitted conditions
Message for the abrogation of coordination	MAC	Used for cancelling the previously sent coordination messages

for the air traffic controller, who has transmitted the original message (the only exclusions are the sent ABI messages).

Note: The LAM does not preclude the requirement for transmitting the technical messages on the integrity of the received data.

ABI Message

According to the OLDI Standard, a notification message is required to be sent to the successive ATS unit. The required notification is an **ABI** message. Note: When the period between the aircraft departure and the time of the aircraft's reaching the coordination point (COP) is short, a preliminary activation message (**PAC**) is transmitted instead of the **ABI** message. **ABI** messages allow as follows:

- to obtain the information on the condition of crossing the FIR boundary of the successive ATS unit by providing the preceding information on boundary crossing;
- to update and obtain the flight plan data, to amend the flight plan in the database;
- to provide the correlation between the track and the flight plan in advance;
- to assess the expected medium-term workload on the air traffic controller of the receiving sector;
- to request the change of the SSR code if necessary.

When amending the estimates, it is allowed to transmit the recurrent ABI message. The ABI message is received automatically, without any actions from the air traffic controller no later than 30 min prior to the aircraft entry into the area of the air traffic controller's responsibility. The time value for the ABI message transmission is established in the two-party agreement individually for every entry point. Then a LAM is transmitted. The received ABI message is automatically associated with the relevant flight plan. If the relevant flight plan is not found, the message is automatically queued for viewing, estimating, and creating a plan on the basis of this ABI message. For further amendments, the activation messages (ACT, PAC) are used. The ABI transmission is always prior to the transmission of the activation messages.

Table 2.2 Message in the flight plan

Message type	Message type, message number	Flight number and SSR code	Flight rules and type of flight	Number of aircraft in a group, aircraft type, and turbulence category	Equipment	Aerodrome of departure and time of departure	Point and flight level of entering aerodrome traffic circle	Flight route	Aerodrome of arrival, total used time, alternate aerodromes	Additional information
Flight plan field No.	3	7	8	9	10	13	14	15	16	18
ABI	+	+	+	+	+	+	+	+	+	+
ACT	+	+	+	+	+	+	+	+	+	+
LAM	+									
PAC	+	+		+		+	+	+	+	+
REV	+	+	+		+	+	+	+	+	+
MAC	+	+				+	+		+	+

The transmission of the ABU shall be interrupted if the activation messages are being transmitted. The ABI message is generated and transmitted to the adjacent control unit automatically if the time remaining before the aircraft leaves the control area is less than the designated (VSP) for transit flights, and for the departing flights—when the actual time of departure (ATD) is entered. Figure 2.38 presents schematics for transmission and receipt of the ABI message. If the flight plan is found in the system (or re-entered) and shows no contradictions with the data, contained in the ABI, when the aircraft enters into the observation aids' coverage area no earlier than 30 min prior to the aircraft's entry into the control area, an aircraft monitoring log is displayed with the flight number of the associated message with the digits in gray.

With that, the "ABI SENT" annunciation is displayed in the "STATUS" field of the daily plan list, of the planning list and the controlled aircraft list. When the response LAM is received, the "LAAM ABI" annunciation is displayed in the above-mentioned lists (i.e., the ABI receipt is confirmed). In case no LAM is received from the adjacent control unit, the "ABI NO LAM" warning is displayed in the "STATUS" field of the daily plan list. Prior to the activation message transmission, the recurrent ABI messages may be sent or some amendments may be made to the following types of data:

– the flight route if the entry point has been changed;
– the aerodrome of departure;
– aircraft type;
– boundary-crossing flight level;
– SSR code in the transfer control point;
– estimated time of passing the entry point is more than the time, established by the agreement;
– any other data, established by the two-party agreement.

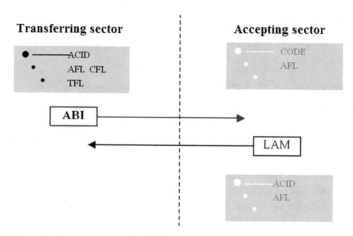

Fig. 2.38 Transmission and receipt of an ABI message

In case the system contains the flight plan, associated with the received ABI message, but the ABI message contains an error, or there are data discrepancies in-between the ABI message and the associated flight plan, the "ABI CORR" warning is displayed in the "STATUS" field of the associated plan in the daily plan list.

ACT and PAC Messages

The activation message provides the main information on the aircraft control transfer in-between the two ATS units in standard conditions. In case there are ATS units equipped with the instrument communication facilities, the activation message is generated and transmitted automatically within the time period, stabled by the agreement [no less than 10 min prior to the ETO of the entry point crossing (VSP)]. The activation procedure can be performed by transmitting either an ACT or a PAC message. 10 min (VSP) prior to the boundary crossing, the system of the transferring sector automatically generates a message, containing the actual data on the conditions of the aircraft's entry into the airspace of the successive control sector along the route with the amendments, entered by the air traffic controller of the transferring sector in relation to the given aircraft. With that:

1. The system shall generate a reminder message for the transferring ATS unit on the required transmission of the associated ACT message to the adjacent control unit by means of the voice communication system), if there is no instrumental interface with the system in accordance with the OLDI Standard.
2. In case there is no instrumental communication, the data on the conditions of the aircraft entry/leave of the control area are transmitted in voice, based on the data, called out manually from the system by means of the control panel inputs.
3. In case the flight is conducted on the non-standard flight level or along a non-standard route, the OLDI allows the air traffic controller of the upper ATS unit to manually transmit the activation message to the lower ATS unit, then this message is called a "Referred Activation Proposal" (RAP) message and is sent instead of the ATC message. If defined by the agreement conditions, the RAP may be also sent in standard conditions of the control transfer of a certain special flight.

The purpose and the aims of the activation message are as follows:

- substitution of the voice communication for transmitting the estimated conditions of crossing the FIR boundary by means of the automatic transmission of the flight data from one ATS unit to the successive unit 10 min (VSP) prior to the control transfer;
- updating the main flight plan data at the receiving ATS unit with the latest information, available 10 min prior to aircraft's passing of the entry point;
- option of correlating the amendments to the conditions of crossing the sector boundary of the adjacent FIR.

Note: If the communication link is inoperative, the ACT and PAC messages are transmitted via the voice communication system.

ACT Message

The system automatically generates and transmits the ACT message, if the transmitted data correspond to the conditions of the agreement on the entry point (standard conditions). Only one message is transmitted. ACT messages are generated and transmitted automatically no less than the established time period (VSP) prior to the aircraft entry to the adjacent automated ATS unit, no associated actions are required from the air traffic controller. If the adjacent control unit is not automated, the ACT message data transmission is performed via the voice communication system.

Displaying the Data at the Transferring ATC Unit

When the ACT message is sent, the "ACT SENT" annunciation is displayed in the "STATUS" field of the daily plan list and of the planning list. The LAM shall be received from adjacent ATS unit. With that, the "LAM ACT" annunciation is displayed in the above-mentioned lists. As soon as the LAM is received in the transferring ATS unit, the ACT message data shall become mandatory for the both ATS units. When the established time period is over, but the LAM has not been received, the "ACT NO LAM" annunciation is displayed. The recurrent ACT message may be sent for the same flight, if the previous messages have been cancelled by the MAC message (see below). During the time period, established by the agreement in between every adjacent ATS unit:

– in the first line of the aircraft monitoring log, the ACT is displayed in yellow digits on the screen background;
– In the "STATUS" field of the daily plan and the plan lists, the ACT reminder message is displayed, requiring that the information shall be sent to the adjacent ATS unit via the voice communication system.

Displaying the Data at the Accepting ATC Unit

When the ACT message is received automatically, the associated line is transferred to the active aircraft section (purple background) of the daily plan list, the "ACT RECV" annunciation is displayed in the "STATUS" field (standing for the received ACT).

The line of the associated flight is displayed in the planning list, and the "ACT RECV" annunciation is displayed in its "STATUS" field, standing for the receipt of the ACT message. In the entry list, the line with the information on the entering aircraft is displayed. The aircraft monitoring log digits turn blue (i.e., the aircraft entry is being planned). For the transferring sector: the symbol of the successive sector (NS—"next sector") is displayed in the aircraft monitoring log, and for the accepting sector, the associated line is displayed in the entry list within the established time period.

The aircraft monitoring log of the accepting air traffic controller after the receipt of the ACT message is provided in Fig. 2.39.

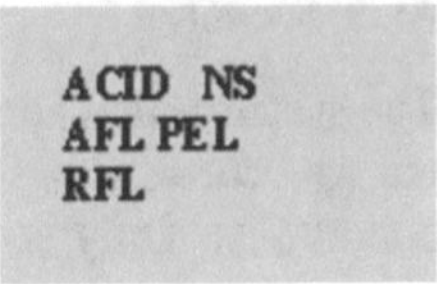

Fig. 2.39 Aircraft
monitoring log of accepting
controller after ACT
message receipt

ACID NS
AFL PEL
RFL

PAC Message

The system automatically generates a PAC activation message, if the transmitted data correlate with the conditions of the agreement on the entry point (standard conditions); however, the time period from the aircraft's takeoff to its passing of the entry point is shorter than the parameter, established for the ACT message transmission.

An agreement is required to include the requested SSR code or route into the PAC message. Prior to the aircraft's departure, an additional PAC message is transmitted, if there are some amendments to the information, contained in the latest sent PAC. As a result, the "PAC SENT" annunciation is displayed in the "STATUS" fields of the daily plan list and the planning list. When the PAC message is received, the data are displayed in the same way as described above for the ACT message receipt.

REV Messages for Specifying the Aircraft Entry Conditions

REV messages are used by the transferring ATS unit for specifying the coordination data, transmitted previously in the associated ACT or PAC message if the accepting ATS unit has remained the same after modifying the flight data. The REV message amends the field values, which have been transmitted earlier in the ACT or PAC messages, if the time, remaining before the aircraft crosses the coordination point, is more than 3 min (VSP); otherwise, the coordination is conducted via the voice communication. The REV message is automatically generated only if the accepting ATS unit has remained the same after the conduct of the latest coordination; in the other case, the MAC message generation and transmission is required, followed by the initiated coordination with the new ATS unit. When the REV message is transmitted, the "REV SENT" annunciation is displayed in the "STATUS" fields of the daily plan list and planning list. The REV message is sent when the amendments are within the standard conditions requirements, established by the agreement. The REV message contains the same information fields as the ACT message. Several REV messages can be sent to the accepting control unit. The REV message is automatically transmitted immediately, when the system of the transferring control unit detects the changes in the entry/release conditions, which have been agreed in advance, which appeared due to the amendments, either entered by the transferring air traffic controller or any other changes, affecting the entry/release conditions. The REV messages are transmitted due to the following changes of the flight performance parameters: estimated time of passing the coordination point (ETO); the change of the coordination point (COP) while the sector remains the same; the transfer flight level (TFL); the SSR code; aircraft equipment operational parameters (only in regard to W and X).

The received REV message is associated with the relevant flight plan in the accepting control unit. The flight plan fields are accordingly amended, and the associated annunciation is generated for the air traffic controller, indicating the changes and the causes of these changes (the same cause is indicated in the MAC message):

- "TFL"—if the transfer flight level has been changed;
- "RTE"—if the route has been changed;
- "CSN"—if the call sign has been changed;
- "CAN"—if the plan has been cancelled;
- "DLY"—if the cause is a delay;
- "HLD"—if the aircraft has been instructed to hold;
- "OTH"—any other cause or if the cause is not defined.

If the agreement requires the indication of the cause to be transmitted, the cause is transmitted via the voice communication system. The REV message can be sent on condition that the aircraft has reached the time and distance values before the coordination point, established in the agreement for every COP, when the PAC message can still be transmitted. If the conditions are standard and the agreement requires no acknowledgment of the amended parameter, the changed fields in the entry list of the aircraft monitoring log turn white for 15 s and then turn to be displayed in their standard way. The system of the accepting control unit confirms the receipt of the REV message by transmitting the LAM, if the changed are within the standard conditions. When the accepting controller receives an ACT, PAC, or REV message, but does not agree with the sector entry conditions, provides the satisfactory sector entry conditions via the voice communication system.

The schematics of the REV message transmission/receipt are provided in Fig. 2.40.

Massage of Abrogation of Coordination (MAC)

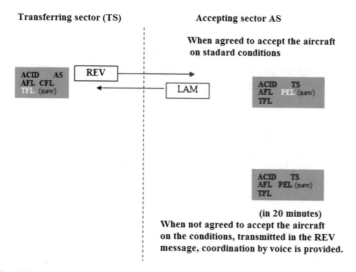

Fig. 2.40 REV message transmission/receipt schematics

The MAC message is used to inform the accepting control unit on the abrogation of the previously performed coordination or notification. This message does not delete the basic flight plan data. The MAC cannot be used to cancel the flight plan.

MAC Transmission

The MAC is automatically transmitted in order to abrogate the flight activation, performed upon the receipt of the ACT or PAC message, in the flowing cases;

- The transferring control unit has received the MAC on the given flight from the adjacent higher-located control center.
- The flight level, at which the boundary crossing is expected, has been changed, causing the change of the originally agreed successive sector for the aircraft acceptance.
- The flight route has changed causing the change of the originally agreed successive sector for the aircraft acceptance.
- The flight plan has been cancelled at the transferring control unit, so further coordination is not required.

The MAC shall also be transmitted, when the coordination on the departing flight, for which the PAC message has been received, is amended.

If the MAC is sent, the notification, activation, and coordination on the flight shall be performed anew with the newly designated control unit. If prescribed in the two-party agreement, the MAC shall contain the reference to the ABI, PAC, or ACT messages for the given flight, which have been sent and acknowledged earlier. The new coordination point (COP) shall be provided, for which the new crossing of the boundary in-between the two adjacent sectors is to be performed. When the MAC is transmitted, the associated line of the daily plan list shifts to the "green" area, and the "MAC SENT" annunciation is displayed in the "STATUS" field of the aircraft list. When the associated LAM is received, the "LAM MAC" annunciation is displayed in the above-mentioned lists and the aircraft monitoring log. In the other case, the "MAC NO LAM" annunciation is displayed. IF according to the agreement, the cause of the specifications is required to be announced, the cause is announced by means of the voice communication system.

MAC Receipt

The LAM message is required to be sent in case there are no discrepancies, which may affect the processing. If the message cannot be associated with a flight plan or some discrepancies, preventing the processing, have been found, the LAM shall not be transmitted and the associated warning shall be displayed. At the accepting sector control unit, when the MAC is received, the abrogation of the preformed coordination which is in the line of the entry list (SIL LIST) is displayed.

Typical Responses for Messages

The OLDI Standard provides several types of responses for various messages. For some messages, only an acknowledgment is required, some require acceptance or declining responses, for others there is a dialog option, when the reply can provide the

alternative suggestion, for which, in its turn, there is also an option of either acceptance or declining response. When the system receives the message, it automatically
transmits an acknowledgment for every message, which requires an acknowledgment only. Other requests or suggestions can require a response from the accepting
ATS unit. The items of such messages, which require coordination, are displayed for
the air traffic controller for acceptance, decline or, in some cases, alternative suggestion. When the alternative suggestion is provided as a response, the former air traffic
controller, in their turn, shall also responds either with a consent or a decline.

The ICAO AIDC Standard

In the mid-1990s, the ICAO have developed Doc 9694, entitled "Manual of Air
Traffic Services Data Link Applications," where, among other issues, the application
of the use of the data links (DL) among the ATS units is described under the name
of "ATS Interfacility Data Communication" (AIDC). As the OLDI Standard, the
AIDC Standard requires transmission of the notification on the aircraft, approaching
the FIR boundary, the coordination of the boundary-crossing conditions as well as
the communication/control transfer by means of the digital exchange of the ATS
messages in a dialog form of notification, coordination, and transfer. As for the basic
procedures, the AIDC messages are very similar to the OLDI messages (although
their names are different). The description of the message types and their contents
are provided in ICAO Doc 4444 Appendix 6 (Revision of 2007). Contrasting to
the OLDI, the AIDC provides no additional messages, such as, the messages for
coordination of military and civil aviation flights. The AIDC Standard is applied
in China, India, North Atlantic and Asian-Pacific areas, Australia, New Zealand,
Indonesia, and many other countries. In the Russian Federation, the AIDC Standard
is not applied so far, although there are agreements between the Russian Federation
and the countries of the associated area, which contribute to its application in the
Far-Eastern regions of the Russian Federation.

2.3.5 Task of Organization of Arriving Traffic Flows—AMAN

The term arrival management (organization of arrival flows) is known in the world
experience (including Europe). This general term includes the process of safe and
economically efficient organization of arriving A/C in order to regulate and streamline
the flow of the A/C arriving at a specified destination aerodrome. The process of
planning the arriving A/C flow is a multi-level process and is connected with different
elements and stages of ATM functioning. This includes organization of airspace use,
arrival and departure route schemes, ATS procedures, planning and coordination of
airspace use, ATS automation means, organization of air traffic flows, etc. The task of
organization of arrival flows starts already on the stage of organization, planning and
coordination of airspace use, continues while organizing air traffic flows on the stages
of strategical, pre-tactical, and tactical planning of airspace use, and is completed

and implemented during ATC process. Thus, the organization and planning of arrival flows is a continuous interstage process of multi-level planning aimed at the stage of ATC (and thus considering limitations and conditions of the actual air situation). Figure 2.41 shows a generalized concept of implementing the task of organization of A/C flow.

Purpose, Tasks, and Types of AMAN There is a wide range of procedures and auxiliary means, different in their complexity, which are used in AMAN. Today, the most completed are the special means (software) which provide not only automatic organization of arriving A/C according to their arrival sequence and optimization of the process, but also information (or recommendations) for air traffic controllers on the organization and regulation of this sequence. Such specialized means of automation are called Arrival Manager (AMAN) [7, 8]. So, the AMAN means (or function) is a means of automation of controller's decisions on organization and regulation of arrival flows.

The purpose of AMAN is to increase efficiency of airspace use and increasing the capacity of the aerodrome area as a result of automation of the processes of organization and regulation of arrival flows which converge into specified (control) points (RWY THR or tracks contingency).

The main tasks of AMAN are: optimal use of RWY by organization of arriving A/C sequence and/or regulation (streamlining) of the A/C flow incoming into the TMA airspace to provide predictability of air situation; decreasing delays, minimal use of orbital delay procedures and vectoring, minimization of negative impact on the environment. The tasks performed by AMAN include the following:

1. organization of arriving A/C sequence for one RWY (in case several RWYs are used—for each RWY individually);
2. estimation of planned time of arrival (ETA) or overflight (ETO) for fixed control points on the arrival route (approach pattern);
3. measurement of arrival A/C flow load for each RWY (in case several independent RWYs are used), aerodrome (for dependent RWYs) or characteristic point which has limitations of capacity (in order to control the arrival intensity/arrival intervals);
4. estimation and implementation of mainly linear delays providing optimal flight path profiles and minimization of orbital holding;
5. organization and regulation of arrival sequence based on specific criteria and limitations concerning delays of arriving A/C in each responsibility zone of area and aerodrome control towers;
6. determining arrival intervals to organize a sequence in accordance with the limitations of the control point capacity;
7. displaying of arrival sequence as well as values of necessary A/C delays (acceleration) through human–machine interface on each workstation in area and aerodrome control towers;
8. organization and regulation of arrival sequence considering RWY limitations as well as departure slots;

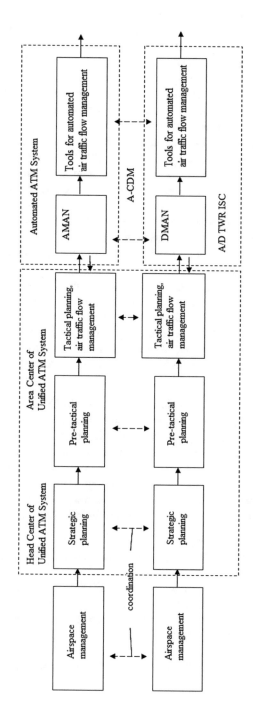

Fig. 2.41 Generalized concept of implementing the task of organization of A/C flow

9. providing control capability for additional criteria when organizing a controlled arrival sequence (minimization of an average delay, increasing the number of the AC controlled, economical efficiency, etc.);

10. providing capability of inputting the planned changes of RWY working direction and their accounting during a specified forecast time interval (VSP), at the beginning of which the previous sequence is stopped so that the latest previous AC can approach before the RWY change, and all the following A/C form a new sequence for the new RWY direction;

11. automated data acquisition on the actual time intervals for individual flight phases of arriving A/C (AMAN statistics acquisition module);

12. reforming arrival sequence in case of manual sequence change or inserting an AC, changing the RWY, inserting departing slots and closing the RWY, prioritizing A/C control;

13. providing capability of servicing all the aerodromes which are included in the AMAN working area with one or several RWYs;

14. providing capability of manual deleting and inserting A/C into the arriving sequence being formed;

15. providing capability of adjusting configuration of functions depending on the parameters chosen and corrected manually through human–machine interface;

16. integration with other automation means for organization of flows (DMAN);

17. as a rule, the use of AMAN is performed by a specified person (e.g., a coordinating controller), and the related information is distributed to the workstations of approach controllers and aerodrome control tower controllers of involved area tower sectors.

In the context of operational procedures, the use of AMAN means is necessary for proper organization of arrival sequence and approach priority in order to increase RWY capacity, considering such factors as separation intervals depending on A/C category of wake turbulence intensity, local limitations and conditions, and RWY occupation for landing. In relation to human–machine interface, arrival sequence is an organized list of A/C passing a fixed controlled point where each A/C has a prescribed (planned) moment of overflight (Target/Required Time Over—RTO/TTO). As a means of automation of controller's decisions, the AMAN means allows:

- to decrease controller's workload;
- to increase economical efficiency of flights by decreasing delays of arriving A/C (primarily in hold areas);
- to rationally use available ATC system resources (RWY construction, control tower organization, aerodrome scheme, etc.);
- to increase the available aerodrome capacity by decreasing A/C intervals;
- to increase RWY capacity when used in arrival mode and mixed mode (arrival/departure);
- to provide the required air traffic safety system by mitigating controllers' peak workload.

The AMAN means can be divided into three main categories:

1. non-specialized regulation tools based on main functions of the flight data processing system (FDPS);
2. specialized means—AMAN basic version;
3. specialized means—AMAN prospective versions (being designed and purposed for future use).

The minimum set of functions of the AMAN basic version is the following: organization of arrival sequence for one chosen control point; organization of an optimal sequence; providing a completed sequence.

Organization of arrival sequence usually starts on the planning stage when the AC position is not fixed yet, and it can move in the sequence within a specified time interval. The following stages of the sequence are more fixed when less change in sequence position allowable or possible. Final completion of the sequence means it is completely fixed when no changes are possible. Forming an optimized sequence is usually performed using a number of criteria which can vary depending on local conditions: even distribution of delays, A/C categories, RWY capacity and limitations, etc. An optimized arrival sequence, as the main result of AMAN functioning, can be provided for the following control points: point of control transfer threshold (point of sector entry); points set to streamline and regulate A/C flows; initial approach fixes (IAF); point of RWY edge overflight; any other prescribed trajectory points.

To implement the formed sequence, it is presented by the human–machine interface with a link to the time axis, data on the planned target time of overflight of the control point and recommendations on the required delay (or acceleration): vectoring, route lengthening, speed change, procedure of delay schemes. In some cases, the AMAN means can generate special recommendations related to the speed values or commands of fine regulation maneuvers in the related approach pattern areas (e.g., using the Point Merge System pattern).

It is significant that, in accordance with the world experience of developing and introducing AMAN, it is NOT a means of:

- full trajectory control;
- locating and eliminating conflicts;
- an intellectual control replacing a controller;
- making final decisions in ATM system whose following is mandatorily required.

EAT can be considered as target time used for A/C control which must necessarily be observed, e.g., by onboard FMS. It can be provided for the area control tower controller as a prescribed condition of entry into the approach area sector of the aerodrome control tower to be implemented by different procedures available (route change, hold, vectoring, speed change).

The sequence number can be determined based on the EAT and ETA.

The outer border of the AMAN procedure functioning is determined by the time of initial ETA available, which is acquired by trajectory calculations. The inner border of the AMAN procedure functioning is determined by the time of the arriving A/C landing.

Operational Concept of AMAN Functioning

Operational concept of AMAN functioning provides the following general scenario (Figs. 2.42 and 2.43):

1. At the distance of 250–400 km from the RWY edge of the destination aerodrome, the aircraft is assumed for monitoring and starts being serviced by the AMAN means (horizon or threshold of AMAN functioning).
2. The AMAN means estimates the preferable A/C arrival time (without limitations).
3. Based on the preferable arrival time and criteria of forming (optimizing) the sequence, the A/C position in the arrival sequence is determined.
4. By using the human–machine interface, AMAN displays the data (warnings and recommendations) which are used by the controller to choose the control commands for A/C regulation and forming the intervals between the A/C in the arrival flow by means or radiotelephony communication.
5. The AC crew follows the controller's commands.

The AMAN input data include: planned data, radar data, regulations and limitations of airspace use, regulations and limitations of flight operations, A/C categories and technical characteristics, meteorological data, data of active (manual) input by human–machine interface from workstations (RWY capacity, separation minimums, intensity of RWY arrivals, intervals of RWY blockage), optimization criteria.

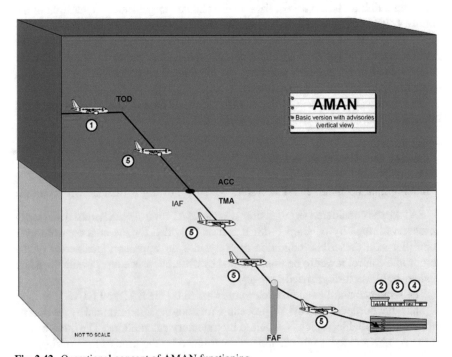

Fig. 2.42 Operational concept of AMAN functioning

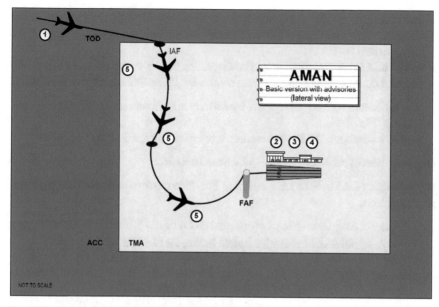

Fig. 2.43 Operational concept of AMAN functioning

The limitations and considered factors include:

- A/C longitudinal separation minimums during controller's approach servicing and aerodrome control servicing;
- RWY arrival intervals;
- separation intervals depending on A/C category of wake turbulence intensity;
- norms of RWY capacity;
- local limitations;
- ATM;
- ATS procedures used;
- RWY limitations and modes of use;
- limitations connected with departure slots when the RWY is used in a mixed mode (arrival/departure).

As a result of AMAN functioning, the following data are displayed on the workstations of ATS system by human–machine interface:

- optimized A/C arrival sequence;
- data presented in the aircraft monitoring log: A/C identification (flight number), sector entry point (direction), category of wake turbulence, necessary delay/acceleration time, priority of servicing, other data (RWY change, aerodrome flights, necessity of hold, etc.);
- target arrival/overflight time;

- recommendations for delay/acceleration time, speed, start time of prescribed maneuvering (when using fine regulation procedures), point and/or time of initial approach fix (IAF);
- results of AMAN functioning can be displayed on the workstations of aerodrome and area control towers and airport services for information and coordination.

The process of forming a sequence by AMAN in the general view can be presented as follows (Fig. 2.44):

Figure 2.45 shows the system elements of providing the AMAN functions.

Organization of AMAN Human–Machine Interface

Organization of AMAN human–machine interface must provide the following display functions:

1. display of estimated time of overflight (ETO) of RWY control point on the time axis of actual arrival sequence linked to the log of every arriving A/C;
2. indication of the RWY for which arrival sequence is formed;
3. display of the actual arrival sequence for the chosen RWY (vertical time axis, arriving A/C logs listed in the order of the formed overflight sequence for the RWY control point and linked to the specified moments of overflight);
4. display of the formed AMAN arrival sequence (uncontrolled/controlled) for the chosen RWY (vertical time axis, arriving A/C logs listed in the order of the formed overflight sequence for the RWY control point and linked to the specified moments of overflight);
5. display of the required time of overflight (RTO) for the RWY control point on the time axis of the formed AMAN arrival sequence linked to the log of every arriving A/C;

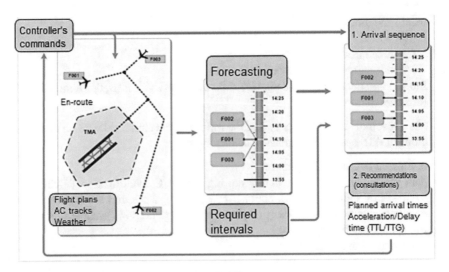

Fig. 2.44 Process of forming a sequence by AMAN

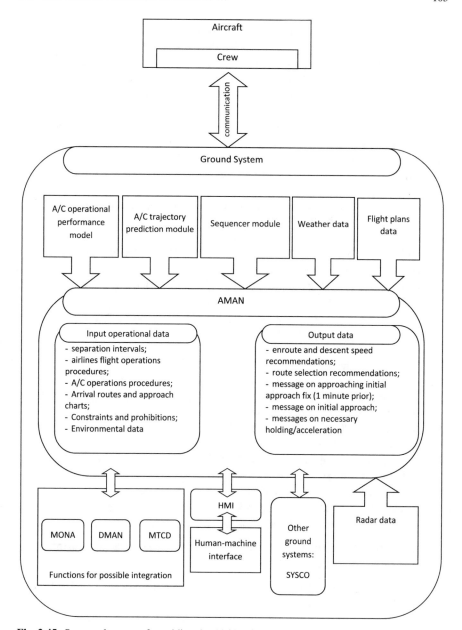

Fig. 2.45 System elements of providing the AMAN functions

6. display of criteria of optimizing the formed AMAN arrival sequence;
7. display of the necessary delays/acceleration (difference between ETO and RTO for the RWY control point) in the log of every arriving AC/the AMAN window/the window of arrival lists/the radar logs;
8. display of the actual number of arriving A/C for the chosen time period in the AMAN window;
9. display of the allowable number of arriving A/C (intensity of arrival/intervals of arrival) for the chosen time period in the AMAN window;
10. display/selection of the fixed/unchanged part of arrival sequence;
11. display of RWY occupation intervals by every arriving A/C of the fixed part of arrival sequence;
12. display of separation intervals between all or specified A/C in the AMAN window/arrival list window;
13. display of the list of added/excluded A/C;
14. color coding of the displayed necessary time of delay in the log of every arriving A/C for the most probable way of its implementation;
15. color coding of the log frame of arriving A/C in the AMAN window for approach direction (entry point into approach area);
16. display of the log of arriving A/C in the AMAN window: A/C identification, WCT, necessary delay time, additional data;
17. display of arrival list window;
18. display of window for arriving A/C separation;
19. display of window of change of A/C position in arrival sequence;
20. AC warning (blinking) which needs manual positioning in the arrival sequence in case of recurrent approach in the fixed part of the sequence;
21. warning for A/C, which are impossible to add into the fixed part of the sequence without its re-estimation;
22. display of the necessary time of overflight of the control point (merge point, IAF, load dosing point) for the chosen arriving;
23. display of time interval of RWY blocking;
24. display of the number in the approach sequence in aircraft monitoring logs for A/C included in the fixed part of the arrival sequence.

The human–machine interface of the AMAN means must implement the following functions of active (manual) data input:

1. choice and display of the chosen RWY control point (outer marker light, inner marker light, RWY edge);
2. choice of the fixed point (merge point, IAF, load dosing point) and display of the RTO for each overflying A/C;
3. choice of RWY for which arriving sequence is formed;
4. choice of optimization criteria for the formed AMAN arrival sequence;
5. choice of allowable number of arriving A/C (intensity of arrival) for the chosen time period in the AMAN window;
6. choice of time interval for RWY blocking;
7. entering priorities for specified arriving A/C;

8. entering additional A/C into the AMAN arrival window (from the list of added/excluded A/C);
9. excluding arriving A/C from the AMAN arrival A/C (adding to the excluded AC list);
10. entering and display of separation intervals between all or specified arriving A/C in the AMAN window/arrival list window;
11. manual correction of the formed arrival sequence by moving the aircraft monitoring log in the AMAN window.

The list of AMAN forms can include the following main elements: AMAN window for approach controller, AMAN window for area controller, AMAN window for coordinating controller, elements of displaying AMAN data in radar monitoring logs at approach controller's workstation, elements of displaying AMAN data in radar monitoring logs at area controller's workstation.

Below is the description of a version of implementing the main HMI AMAN element—the AMAN approach controller's window.

The approach controller's AMAN window (Fig. 2.46) consists of the top panel, sequence field, warnings text field, and settings and functions bar.

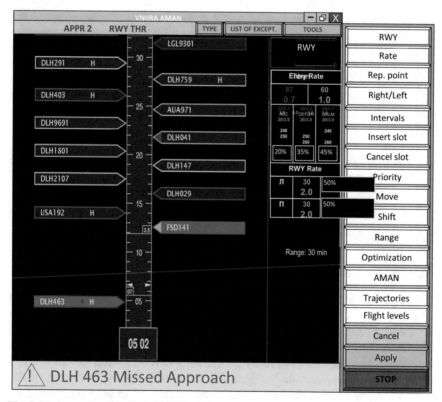

Fig. 2.46 Approach controller's AMAN window with settings and functions panel expanded

The AMAN window top panel has a title line and the line for designation of workplace and RWY control point. In the title line, there are the following buttons: Minimize, Expand, and Close. The line for designation of workplace and RWY control point, there are the buttons: View, Excluded list, and Tools.

The sequence field of the AMAN approach window is a two-sided vertical time scale which allows to display the arrival sequence considering separation intervals for two parallel RWYs.

Positioning of arriving A/C logs depends on the planned landing RWY (left/right), time of overflight of the RWY control point for this A/C, as well as sequence of arriving A/C for final approach. The right top area of the sequence field has working RWY designation, value of arrival intensity and transfer flight levels in each entry point, RWY use intensity, range of time scale displayed. The bottom part of the time scale has the current time field below the horizontal red line.

The tool bar is to choose the working RWY (RWY button), control point (control point button), time scale range (range), optimization criteria (optimization), entry flight level for the aerodrome area for arriving AC in each entry point (flight levels), default approach patterns for all and specified AC (trajectories), values of arrival intensity (intensity), minimum separation intervals (intervals), moving the log from one parallel RWY to another (right/left), inserting or cancelling the slot (insert slot/cancel slot), moving the logs along the time axis without change (move), and with sequence change (change), prioritizing (priority), start or update of sequence estimation (AMAN), stop of AMAN (STOP), as well as the buttons Cancel and Apply.

The AMAN warning text field is immediately below the sequence field. In case there are no warnings, the field is minimized; in case there is a warning indication, the field expands and its left area displays the warning sign which precedes the warning text.

The log for each arriving A/C contains (Fig. 2.47) the A/C identification, WTC, prescribed transfer flight level in assuming control (dosing) point, indicator and value of required delay/acceleration in minutes.

2.3.6 Aid for Managing Departing Aircraft Flow, DMAN

Purpose, Aims, and Tasks of DMAN

The Departure Manager (DMAN) is a planning aid, which allows increasing the efficiency of managing and planning the flow of departing aircraft in order to optimize the use of the airport resources and available ATC capacity [12]. The purpose of the DMAN functioning is to plan the order of departing aircraft in order to provide the optimum runway traffic capacity (effectivity), to decrease the aircraft queuing before lineup, and to provide the necessary information for the concerned parties (airline companies, aircraft flight crews, airport services; ground support services, ATS units).

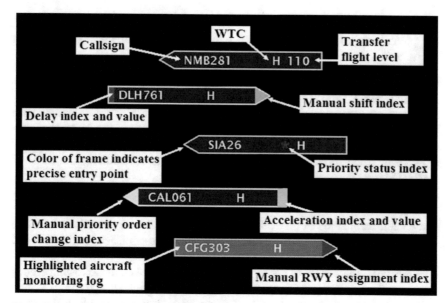

Fig. 2.47 Departing aircraft monitoring log

By means of the HMI, the sequence of departing aircraft is displayed as a list of departing flights in the DMAN planning horizon with the indication of the calculated values of the target take-off time (TTOT), target engine star-up time (TSAT), and the relevant estimated time of wheel chocks removal (TOBT—Target Block-Off Time). The above-mentioned information is displayed at the workstations of the specially assigned personnel of the apron management support department (if available), departure dispatch management station ("Delivery"), the taxing control unit (TCU), the line-up control unit (LCU), as well as at the workstations of the aerodrome flight dispatch supervisor (AFDS) and the coordinating controller (CC). The main functioning task of the DMAN is to determine the TTOT, the corresponding TSAT or the TOBT, taking into account various constrains and preferences.

The functional tasks, which are solved by the DMAN, are as follows:

- automatic calculation of the time intervals, available for planning take-off operations;
- automatic calculation of the time intervals, restricting the performance of take-off operations due to turbulence conditions and separation requirements;
- calculation of the time, required for aircraft delay at takeoff;
- calculation of the recommended time for engine start;
- optimization of the departure sequence, based on the following criteria:

  - maximizing the runway capacity;
  - minimizing taxiing delays (delays in standby areas and during taxiing);

- maximizing the number of departures (in accordance with the airline companies' preferences and in order to improve the traffic management in regard to slot times).

- automatic calculation of the allowed time (distance) between flights in the landing–take-off and take-off–take-off modes;
- automatic message exchange, based on the information slow of interrelation between the automated ATC systems and automated system of technological process management;
- automatic calculation of time of aircraft transfer to the ATS unit control;
- calculation of the recommended take-off time;
- automatic calculation of time intervals, available for planning the take-off operations;
- automatic calculation of time intervals, limiting the take-off operations, both due to turbulence conditions and for compliance with the established longitudinal and vertical separation limitations;
- automatic re-calculation of the departure sequence, based on the following events: runway change, change of the parking stand (or passenger boarding apron), failure or delay when performing the air traffic controllers' instructions; blocked runway (RWY), designated departure priority;
- distribution of the recommended time intervals among the control units;
- statistical data acquisition (taxiing time, RWY occupation time, time for lineup, etc.).

DMAN Operational Concept

The functionality of the DMAN is based on the planning and optimization, which is dependent of the events. The events mainly include the instructions and clearances, issued by air traffic controllers. Such sequence of the events and the associated time intervals in-between these events are the basis for the departing aircraft movement model, created by the DMAN. The established types of ATS units'/ground support services' instructions include:

- clearance for chocks removal;
- clearance for towing/pushback;
- clearance for engine start-up;
- clearance for taxiing;
- clearance for lineup;
- clearance for takeoff.

The established time intervals in-between the above-mentioned events correspond to the following transition steps in-between the events: towing/pushback, engine start-up; taxiing to lineup; taxiing to the RWY. The de-icing period is taken into account by extending the time interval for taxiing.

When forming the departure sequence, the events, which may trigger its re-calculation, also include:

- operational change of the take-off RWY;
- change of parking stand/apron;
- failure or delay when performing the controller's instructions;
- blocked (closed) RWY;
- deviations from the calculated (model) values of transition time in-between the events;
- request for de-icing;
- establishing the departure servicing priority.

When forming the departure sequence, the following strict limitations shall be taken into account:

- take-off intervals, depending on the intensity of the vortex wake turbulence (WTC);
- departing aircraft's RWY occupation time;
- longitudinal separation in-between the departing aircraft after takeoffs, established for the aerodrome control servicing;
- RWY's positions and the operational models of their use;
- time intervals when the RWY are closed (blocked);
- arrival time for landing aircraft;
- RWY capacity values;
- meteorological conditions;
- de-icing procedures;
- aircraft inflight speed categories;
- SIDs.

The process of forming departure sequence by means of DMAN is based on the following information:

- the operational ATS situation at the aerodrome (the presence of the arriving aircraft, RWY use limitations and RWY condition, RWY capacity, meteorological conditions);
- limitations of the designated slot-time intervals (if there is a procedure for their use);
- aircraft distribution among the parking stands;
- taxiing charts and associated time intervals;
- local constraints, specific for each individual aerodrome.

When planning a natural departure sequence (taking into account only strict limitations), the ATFM plan information and calculated (on the basis of the taking-off aircraft movement model) time values—takeoff, engine start-up, start of towing/pushback; taxiing—are taken as a basis.

If there are procedures for slot-time intervals' use (the airport slot times) and there are requests for preferable departure time slots from airline companies, the DMAN takes these data into account when calculating the departure sequence as "soft" limitations, i.e., they are complied with only when the strict limitations have been complied with. The DMAN algorithm of calculating the optimized departure sequence is based on such criteria as maximum use of the RWY capacity, minimum number of delays during taxiing and prior take-off holding point; maximum correspondence to the target take-off time, provided by the airline companies or ATFM units (correlation to slot-time intervals).

Besides the requirement for minimization of the departure sequence amendments in order to minimize the controller's workload and other criteria are taken into account as well as the option of the air traffic controller's manual override of the DMAN process of generating the departure sequence is ensured. Part of the departure sequence, comprising the aircraft flights, which have already started the taxiing to the take-off holding point, is fixed by the DMAN; and the events, which trigger the departure sequence re-calculation, can include only the active inputs from the relevant workstations of the ATS unit, which provides the aerodrome ATC service. Another important function of the DMAN is the generation of the departure sequence with the mode of the RWY mixed-use mode, which takes into account the time intervals when the RWY is occupied by the arriving aircraft and automatically coordinates the departure slot-time intervals within the mixed traffic flow, when integrated with the AMAN.

The most efficient use of the DMAN and its integration with the other systems are provided by the system of Airport Collaborative Decision Making (A-CDM) [10, 11].

The algorithm of DMAN interaction with the external systems is provided in Fig. 2.48.

All the incoming data, required for the correct operation of the DMAN, are classified in accordance with the following information:

- steps of derating aircraft servicing (cargo loading/passenger boarding, doors' locking, towing/pushback, taxiing, take-off holding, lineup, takeoff);
- FM plan values and calculated values of transition from one of the above-mentioned servicing steps to the other;
- data on the aircraft positions at the aerodrome operational area;
- information on the arrival and departure slot-time intervals as well as/or on the other constraints for takeoff.

The main source of the data for the DMAN operation are the flight data processing system (FDPS) and the airport automated system of technological process management. The A-SMGCS system can also provide additional information on the aircraft trajectory for the purpose of amending the departure sequence in accordance with the actual aircraft movement.

The required data for the DMAN operation include:

- aircraft call sign;
- flight plan index (identification number, used for the correlation of with the track data);
- aircraft type in accordance with the CAO and IATA ratings;
- aircraft category with regard to the vortex wake turbulence (WTC);
- aircraft category with regard to speed;
- used SID routes;
- TOBT;

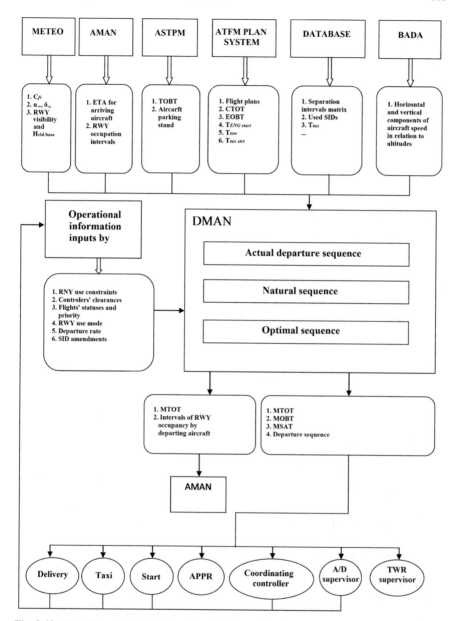

Fig. 2.48 Algorithm of DAMN interaction with external systems

- actual time of chocks removal;
- estimated take-off time;
- applicable time intervals in-between takeoffs and separations in the aerodrome area;
- established taxi routes for the departing aircraft;
- estimated taxiing time in accordance with the routes and aircraft types;
- used direction of the RWY for takeoff;
- actual and forecast meteorological information, data on the take-off RWY condition;
- estimated landing time and the time intervals when the RWY is occupied by the arriving aircraft;
- flight performance parameters of the departing aircraft;
- designated departure slot-time intervals (if available);
- aircraft parking positions;
- RWYs assigned for takeoffs;
- actual time of engine start-ups;
- active amendments to the departure sequence, entered from the controllers' workstations;
- information on controllers' issued clearances and instructions;
- information on the aircraft's reaching of the designated point (holding point, de-icing area, etc.);
- actual aircraft take-off time;
- information the aircraft blocking;
- value for the chosen take-off rate (departing traffic flow capacity).

A generalized algorithm of the DMAN functionality is provided in Fig. 2.49.

DMAN HMI Architecture

The DMAN HMI architecture provides the following functionality:

- indication of the information on the aircraft readiness for takeoff; active (manual) inputting of the data on designated events of the steps;
- servicing planning (issuing of the controllers' clearances, control transfer, reaching the designated points at the area of maneuvering, servicing priority, start of taxiing, start of de-icing, etc.);
- active amendments to the departure sequence, made by the responsible persons of the ATS unit;
- active inputting and operational amending of the hard constraints: take-off intervals, aircraft flight speed categories and take-off routes (SIDs), time of the RWY occupancy, RWY configuration and modes of use, separation intervals, meteorological conditions (minima, required de-icing), intervals of RWY blocking; servicing priority;
- displaying and annunciation of the monitored planned scenario of the departing aircraft servicing;

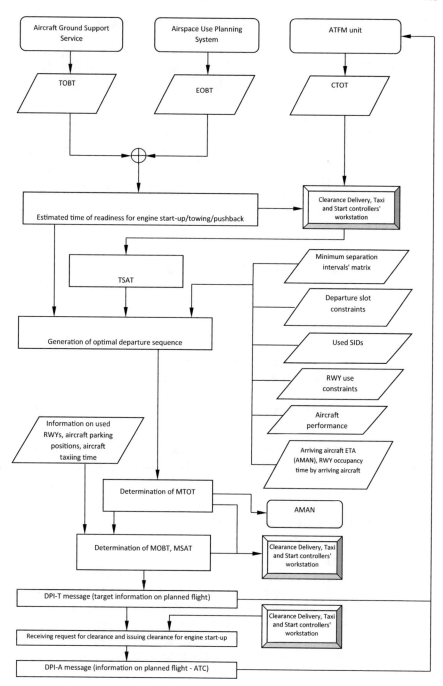

Fig. 2.49 Generalized algorithm of DMAN functioning

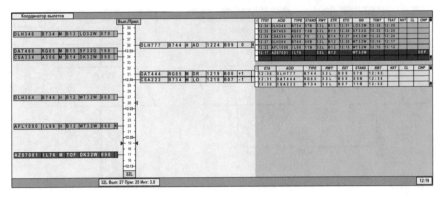

Fig. 2.50 Example of information, displayed in coordinating controller's window for planning arrival and departure sequence

- displaying the departure sequence within the range of the operational horizon of the DMAN planning, including the display of the information on the arriving aircraft.

The main displayed forms of the DMAN HMI include:

- aerodrome charts;
- windows with lists (for departure, holding, etc.);
- window with departure sequence (in form of aircraft monitoring logs, linked to the time scale);
- window for planning the arrival and departure sequence for the coordinating controller.

Figure 2.50 presents the example of the information, displayed in the window for planning the arrival and departure sequence of the coordinating controller.

2.3.7   AMAN/DMAN Integration

The AMAN and DMAN aids can be merged into one integrated functional AMAN/DMAN unit. In case the RWY is used in the mixed arrival/departure mode, the AMAN operational use requires taking into account the established constraints of the RWY use at the aerodrome, serviced with the help of the AMAN aid, and on the departure slot-time intervals. The DMAN operational use requires taking into account the arrival slot-time intervals and the time intervals, available for departures. The following functional tasks of the interaction of the DMAN and AMAN aids can be distinguished:

- the data transmission from AMAN to DMAN on the designated constraints of the RWY use due to its occupancy with the arriving aircraft;

– arrival's rate coordination and the coordination of intervals for departures between the AMAN and DMAN.

The departures, which are planned for the same RWY, which is used for arrivals, can be taken into account in the arrival sequence in form of departure slot-time intervals.

In case DMAN is used as a module, merged with AMAN, when the optimal target take-off time is assigned for the RWY, used for the mixed mode operations (arrival/departure mode), the following issues shall be provided:

– use of the intervals in-between the arriving aircraft in the arriving traffic flow in order to improve the efficiency of the RWY use with the minimum delays for the arriving aircraft;
– calculation of the TTOT, taking into account the time for lineup in order to ensure the preciseness of the correlation with the available time intervals in-between the arrivals;
– taking into account the arrival slot-time intervals, provided by airline companies or the ATM unit;
– combining the departures in order to maximize the departing traffic flow capacity (the so-called flexible rate);
– managing and optimization of the TTOT for the departing aircraft in case AMAN provides the updated (specified) data on the arriving aircraft's arrival times;
– mutual display of the arrival/departure sequence at the involved workstations;
– taking into account the intervals, when the RWY is occupied by the arriving aircraft, and the constraints, associated with the possible go-around in order to provide a non-conflict departure sequence;
– TTOT re-calculation in case the departure sequence has been changed manually or automatically due to the updated arrival sequence.

The standard procedure of including the departing aircraft into the arriving traffic flow implied the extension of the arrival time interval.

The alternative method is the use of departure slot-time intervals.

The departure slot time for the given RWY can be fixed in relation to the exact aircraft in the arrival sequence (i.e., prior to it and after it) or by means of setting the start time of the time slot interval. The slot value is provided by the separation in-between the departing aircraft and both the two arriving aircraft prior to it and after it.

The available options also include the manual slot-time cancelling and slot-time intervals' duration. The start of the slot-time interval of the RWY closing can also be changed.

2.4 Evaluation of Time, Spent by ATC Controller on Defining Values of Minimum Intervals for the Horizontal Separation by Means of Automated ATC Systems

The key objective of the air traffic control services is to prevent the aircraft collision. The main method of its achievement by the ATC controller is the separation, which means the vertical, longitudinal, and lateral distribution of aircraft in the airspace at the required intervals. With that, an important part of the process of decision making and choosing the ATC actions is the minimum separation values and the controller is responsible for maintaining these values, which are the main monitored factor ensuring the safety of air traffic operations. The minimum separation intervals values in the airspace of the Russian Federation depend on multiple factors: the type of air traffic services, the altitude layer of the airspace, the availability and type of available observation systems, the flight direction, the relative positions and trajectories of the aircraft, and even the information on which ministry the aircraft belong to. The determination of the minimum separation values by ATC controllers is the most important part of their activities. The controllers solve this task with the help of the received information on the positions, relative positions movement parameters of the aircraft. Further, on the basis of the evaluation and analysis of this information, the controller shall choose the relevant minimum separation type values (longitudinal, lateral, vertical), none of these values shall be violated. Quick and precise determination of these values is especially important, when the ATC controller decides on the actions to resolve the conflict. It is evident that under the condition of time deficiency, the quicker the controllers decide on the relevant types and values of the minimum separation, the more time is left for their own decision making and acting on the conflict resolution, which leads to the risk mitigation of performing erroneous actions and which is important from the viewpoint of STCA occurrence, when the time remaining before the possible violation of the established requirements—minimum separation values—is less than 80–120 s. The controller's actions are characterized by the speed of reaction and reliability. The main criterion of the speed of reaction when resolving the conflict is the time, spent on solving the ATC tasks, i.e., in this case, the time span from the moment of the controller's reacting on the incoming signal about the conflict occurrence till the time when the conflicting aircraft reach such airspace positions, which prevents their mutual approach for less distance than the established separation minima. Let us evaluate this τ_{on} time on the basis of information theory approaches. From the viewpoint of the information theory, this time is in direct ratio to the amount of the processed information. It is important to take into account that different types of information are processed at the different speed. Consequently, the expression for the estimation can be written in the following way [13]:

$$\tau_{on} = \alpha + \sum_{i=1}^{k} H_i / \vartheta_i \qquad (2.1),)$$

where H_i is the amount of the information of the i type, processed by the operator, ϑ_i is the speed of the i-type information processing, the usual speed for the information processing is 2–4 bits per second; α is the lag phase time, i.e., the time interval from the moment of the signal occurrence to the time of the operator's reaction, $\alpha = 0.2–0.6$ s.

The amount of information, according to Shannon's formula [14], is:

$$H = -\sum_{i=1}^{n} p_i \log_2 p_i$$

where H is the information amount; n is the number of the possible events; p_i probabilities of occurrences of individual events;

For the events with the same probability of occurrence, Hartley's formula is applied [14]:

$$H = \log_2 n, \qquad (2.2)$$

In the case under study, the information, processed by the controller when determining the horizontal (longitudinal and lateral) separation requirements with the use of the observation ATS system and the time, required by the controller to solve this task are considered. Provided that the horizontal separation requirements have the same probability of occurrence, the amount of the information, processed by the controller, H_{contr}, when determining the horizontal separation limits with the help of the ATS observation system (ATS SS) in accordance with (3.2), is:

$$H_{contr} = \log_2 n, \qquad (2.3)$$

where n is the number of the probable horizontal separation limits with the use of the ATS SS.

The time, required by the controller to determine the horizontal separation limits with the use of the ATS SS, τ_{contr} is defined as:

$$\tau_{contr} = \alpha + H_{contr} / \vartheta. \qquad (2.4)$$

When determining the τ_{contr}, let us consider the two extreme cases, when the time of determining the separation requirements is either the maximum or the minimum:

- when the lag phase time is the minimum of $\alpha = 0.2$ s and the speed of the information processing is the maximum of $\vartheta = 4$ bit per second;
- when the lag phase time is the maximum of $\alpha = 0.6$ s and the speed of the information processing is the minimum of $\vartheta = 2$ bit per second.

Let us determine the time, required by the controller for the determine the horizontal separation requirements with the use of the ATS SS, applied in the airspace of the Russian Federation in case of by the reginal control servicing (RCS) and the approach control servicing (ACS) [15].

In order to evaluate the information, processed by the controller when determining the horizontal separation standards with the use of the ATS SS, we can write the following formula:

$$H_{RUS} = \log_2 n_{RUS}, \tag{2.5}$$

where n_{RUS} stands for the number of values of the horizontal separation standards with the use of the ATS SS in the airspace of the Russian Federation (Tables 2.3 and 2.4) [15].

Where ATS SS is the ATS observation system; the automated ATC system is the automated air traffic control system; RCS is the regional control servicing; ACS is the approach control servicing; ADS-B is the automatic dependent surveillance broadcast; IFR is instrument flight rules, VFR is visual flight rules; SFL stands for "same flight level"; SDC of POFL is same direction-crossing of the flight level previously occupied by another aircraft; CC of POFL is the counter-crossing of the flight level previously occupied by another aircraft; CR stands for crossing routes; SDR stands for same direction routes; V_{vert} stands for aircraft vertical speed.

The horizontal separation limits for the airspace of the Russian Federation are defined as all the possible combinations of the longitudinal and lateral separation

Table 2.3 Characteristics of the separation

		IFR/IFR and IFR/VFR					
		SFL		CC of POFL			SDC of POFL
		SDR	CR	ATS routes	State aviation aircraft flight routes		
					$V_{vert} < 10$ mps	$V_{vert} \geq 10$ mps	
RCS	ATS SS + automated ATC system/integrated automation system/ADS-B (km)	20	25	30	60	30	10
	ATS SS (km)	30	40	30			20
ACS	ATS SS + automated ATC system/integrated automation system/ADS-B (km)	10	20	20			10
	ATS SS (km)	20	30	30			20

Table 2.4 Separation standards

Ground		Air	
		IFR/IFR and IFR/VFR	
		CC of POFL	SDC of POFL
RCS	ATS SS + automated ATC system/integrated automation system/ADS-B (km)	10	10
	ATS SS (km)	10	10
ACS	ATS SS + automated ATC system/integrated automation system/ADS-B (km)	10	6
	ATS SS (km)	10	10

limits and are provided in Table 2.4. To evaluate the time, required by the air traffic controller for determining the horizontal separation limits with the use of the ATS SS, we can write the following formula:

$$\tau_{RUS} = \alpha + H_{RUS}/\vartheta = \alpha + \frac{\log_2 n_{RUS}}{\vartheta}, \tag{2.6}$$

The benchmark data and the resulting evaluations are provided in Table 2.5.

To compare, let us determine the time, required by the controller to determine the radar-aided horizontal separation limits (i.e., with the use of the ATS SS), described in ICAO Doc 4444 [3].

The amount of the information, processed by the controller, when determining the radar-aided horizontal separation limitation H_{ICAO} in accordance with ICAO Doc 4444 is as follows:

$$H_{ICAO} = \log_2 n_{ICAO}, \tag{2.7}$$

where n_{ICAO} stands for the number of radar-aided horizontal separation limits in accordance with ICAO Doc 4444.

Table 2.5 Source data for the transfer of information

	RCS	ACS	RCS + ACS
n_{RUS}	12	8	20
H_{RUS}	3.58 bits	3 bits	4.3 bits
τ_{RUS} with $\alpha = 0.2$ s; $\vartheta = 4$ bits per second	1.1 s	0.95 s	1.28 s
τ_{RUS} with $\alpha = 0.6$ s; $\vartheta = 2$ bits per second	2.39 s	2.1 s	2.75 s

As in accordance with the item 8.7.3.4 and item 8.7.3.2 (a) [3], $n_{ICAO} = 2$, then $H_{ICAO} = 1$ bit. And the τ_{ICAO} time, required by the controller to determine the radar-aided horizontal separation limits accordance with ICAO Doc 4444, determined by the expression:

$$\tau_{ICAO} = \alpha + H_{ICAO}/\vartheta = \alpha + \frac{\log_2 n_{ICAO}}{\vartheta}, \qquad (2.8)$$

has the following values:

$\tau_{ICAO} = 0.45$ s with $a = 0.2$ s, $\vartheta = 4$ bits per second; $\tau_{ICAO} = 1.1$ s with $a = 0.6$ s, $\vartheta = 2$ bits per second.

Let us write down the expression for comparing the time, required by the controller for determining the horizontal separation limits with the use of the ATS SS, applied in the airspace of the Russian Federation, and the radar-aided horizontal separation limits in accordance with ICAO Doc 4444:

$$\frac{\tau_{RUS}}{\tau_{ICAO}} = \frac{\alpha + H_{RUS}/\vartheta}{\alpha + H_{ICAO}/\vartheta} = \frac{\alpha * \vartheta + \log_2 n_{RUS}}{\alpha * \vartheta + \log_2 n_{ICAO}}, \qquad (2.9)$$

The results of the comparison are provided in Table 2.6.

Time, required by the air traffic controller for determination of the radar-aided horizontal separation limits in accordance with ICAO Doc 4444 [3], is approximately two times smaller than the time required by the controller for determining the horizontal separation limits with the use of the ATS SS, applied in the airspace of the Russian Federation. Thus, it is deemed possible to mitigate the number of the possible variants of the situations, for which the associated values of minimum separation intervals are established. For example, the horizontal separation limits in accordance with the ICAO Doc 4444 [3] can be applied, which will decrease the time, spent by the ATC controller on the conflict resolution, and will increase the flight safety.

Section Review

1. The difference between the modern automated ATC systems and the automated ATC systems of the previous generation.
2. What integrated complexes and subsystems comprise the modern automated ATC systems?
3. Which basic processes shall be automated in the ATC ISC?
4. Purpose of the ATC ISC.

Table 2.6 The results of the comparison

	RCS	ACS	RCS + ACS
$\dfrac{\tau_{RUS}}{\tau_{ICAO}}$ with $a = 0.2$ s; $\vartheta = 4$ bits per second	2.4	2.1	2.84
$\dfrac{\tau_{RUS}}{\tau_{ICAO}}$ with $a = 0.6$ s; $\vartheta = 2$ bits per second	2.17	1.9	2.5

5. Basic functions of information processing for the ATC needs, realized in the automated ATC ISC.
6. Which types of information are displayed at the automated workstation of the automated ATC systems for the controller?
7. Purpose of aircraft monitoring logs on the air situation display of the ATC ISC.
8. Types of displaying the aircraft monitoring logs on the air situation display of the ATC ISC.
9. What is 4-D trajectory?
10. On the basis of which data the 4-D trajectory is calculated?
11. Name several functions, with which the TP function interacts?
12. Purpose of the MONA function.
13. Which data arė required for the MONA function operation?
14. Which warnings are generated by the MONA function?
15. Which reminder messages are generated by the MONA function?
16. Purpose of the MTCD function.
17. Purpose of the PPD window of the MTCD function.
18. Purpose of the VAW of the MTCD function.
19. Purpose of the OLDI Standard of EUROCONTROL.
20. Purpose, aims, and tasks of AMAN.
21. Purpose, aims, and tasks of DMAN.

References

1. EUROCONTROL Specification for trajectory prediction. Revision 1.0, 15.07.2010
2. EUROCONTROL Specification for Monitoring Aids No. 1.0, 15.07.2010
3. ICAO Doc 4444 ATM/501 Procedures for Air Navigation Services. Air Traffic Management, Edition 15, 2007
4. ICAO Doc 9694-AN/955 Manual of Air Traffic Services Data Link Applications. Edition 2, 1999
5. System Supported Coordination (SYSCO) Implementation Guidelines Edition. 2.0, 18.03.2011
6. EUROCONTROL Specification For On-Line Data Interchange (OLDI) Edition 4.2, Dec 2010
7. AMAN Status Review 2009, EUROCONTROL—Ed. 0.2, Feb 2010
8. AMAN Status Review 2010, EUROCONTROL—Ed. 0.1, Dec 2010
9. Global Air Traffic Management Operational Concept. ICAO doc 9854
10. Airport CDM Implementation: The manual. European Organization for the Safety of Air Navigation (EUROCONTROL), Brussels (2010)
11. Action Plan—Airport CDM Implementation. No. 1.0, CFMU & EUROCONTROL AOE, 09.2008
12. DMAN Operational Concept Development. EUROCONTROL document, Airport CDM Task Force 12, Edition Number 0.4. EUROCONTROL HQ, Brussels, 25.01.2006
13. Ayupov VA, Kupin VV, Playsovskikh AP (2012) Evaluation of span time of the air traffic controller on determining the required minimum intervals of horizontal separation. Scientific Journal "Nauchny Vestnik". Moscow State Technical University of Civil Aviation
14. Kotik MA (1978) Engineer psychology course. Valgus, Tallinn
15. Federal Aviation Regulations "Air Traffic Management in the Russian Federation". Approved by Decree of Ministry of Transport of the Russian Federation, dated 25/11/2011, No. 293

Chapter 3
Purpose and Comparative Study of the Controllers of ATC Automated System Simulators

3.1 Main Directions and Trends of the Simulation Training Devices (STD) Development

One of the topical tasks within the air traffic management system is to improve the working efficiency of the controllers of the air traffic control automated system (ATC AS) as the most important integral part of the flight management system [1]. One of the most important factors of the controllers' working efficiency improvement is their level of professional training, which is inseparable from the use of the Simulation Training Devices (STDs). The process of the professional training assumes receiving of knowledge, building up of practical experience, and transforming of these elements into skills [1]. The relevance of the efficient ATC controllers training is defined by the character of their functional responsibilities during the flight control activities. The integral assessment of the ATC controllers' activities depends on the gained practical experience and skills, on the decision speed and precision that are realized by the controller through the commands and specialized step-by-step technology aimed at the flight safety provision during the air traffic control activities. The responsible actions of the controller in abnormal conditions, emergencies, assistance to the flight crew in in-flight emergencies shall be very well practiced and require very little time for the relevant decision making. The combination of the qualities, which define the current eligibility of the controller to be able to efficiently perform his/her duties [2], shall be continuously improved and monitored with the help of the specialized training aids for the controllers training. The set of requirements [3] to the theoretical knowledge and practical abilities define the methodology of an integral approach to the buildup and organization of the controllers' training aids.

The main components of the training aids are the automated training system (AutoTS) for theoretical training and Simulation Training Devices (STDs) for practical training of the controllers. The particular requirements to the theoretical knowledge and practical abilities of the ATC controllers are the main base that defines the philosophy of the AutoTS and practical training of STDs' buildup.

© Springer Nature Singapore Pte Ltd. 2020
Bestugin A.R. et al., *Air Traffic Control Automated Systems*,
Springer Aerospace Technology, https://doi.org/10.1007/978-981-13-9386-0_3

The main tasks of the AutoTS are the following:

– theoretical training and preparatory practical simulation training of the controllers following the yearly training plans for the simulators flight control trainings;
– self-study with the automated control including a set of AutoTS software for the specific aerodrome;
– assessment of the level of theoretical training of the controllers;
– registering of the results' training-level assessment.

The STDs of the ATC AS controllers are designed for theoretical training, practical training, and conversion training of the ATC AS controllers in order to maintain the professional skills, arrange the training on the in-flight emergencies and abnormal situations, practice technological procedures including those connected to the change in the airspace structure and the implementation of new methods for airspace management, as well as new methods for air traffic management. The STDs of the ATC AS controllers shall provide the simulation of execution of the main technological processes of planning of the airspace use and air traffic control [3] that are realized at the controllers' workstations at the area centers, aerodrome centers, and aerodrome control towers of the specific aerodromes.

With the help of the STDs of the ATC AS controllers, the following main tasks are solved:

– practicing of the procedures for the flight control within a specific aerodrome or ATC center;
– building up of the skills to work with the radio-technical flight control aids, ATC AS, communication means;
– training in defining current and potential conflicts and providing flight safety;
– practicing of skills required for cooperation between the controllers of adjacent control sectors;
– practicing of emergency actions and actions following in-flight emergencies;
– practicing of actions during the aerodrome emergencies;
– practicing of skills to conduct monitoring after visual detection of the aircraft on the glide slope, monitoring of the direction to the runway, flight altitude, position of the landing gear extended, leveling and level maintaining heights, landing, braking and vacating processes, as well as entering of the runway by the aircraft for takeoff;
– practicing the skills to manage the lighting equipment of the specific aerodrome;
– practicing of skills to solve the issues of aircraft breaking down on the runway or runway excursion;
– practicing the skills to monitor the meteorological and ornithological conditions at the aerodrome and within the control sectors.

The main components of the ATC AS controllers training technical aids are AutoTSs for training and retention of theoretical knowledge and normative documentation, and STDs to allow for practical training of the controllers. The main trends of the STDs development can be generalized to the following provisions [4, 5].

1. The principles of the STD buildup are based on the use of the real workstations of the ATC automated systems under development or already implemented as a students' workstations with the human–machine interface of a real ATC system.
2. The STDs are designed based on the open architecture principles (the number of the workstations can vary depending on the structure of the simulated airspace).
3. The leading national and foreign companies implement the synthesis and voice information detection software in order to provide automation for the maintenance staff functions. At that, the operator pilots are not excluded from the team (when complex exercises are played with emergencies and high level of the air traffic intensity, their functional capabilities are fully used). Finally, this allows reducing (but not functionally excluding) the manpower of the operator pilots.
4. A significant advance in terms of ergonomics of presenting the simulated information and improving the cooperation interfaces with the personal computer (PC) at the AWS of the operator pilots; and because of this, the number of the aircraft under control of one operator pilot significantly increases—up to 24 aircraft.
5. The human–machine interfaces are mainly based on the use of the multiple window technology with the possibility to simultaneously open several windows with "attachments".
6. For the STDs' buildup, the universal equipment of general use is utilized (but of a high technology level).
7. As means for the radiocommunication simulation (the communication and control panel at the AWS of the training controllers and of the maintenance staff), the following is used:

   – panel PCs with a set of multimedia technologies;
   – touch screens with the PC.

   At that, the radio voice communication and intercom means are based on the dedicated LAN.
8. Much attention is given to providing for simulation of a vast set of emergencies, in-flight emergencies, and traffic conflicts.

3.2 Controllers STDs Characteristics

3.2.1 Integral System Simulator

Based on the multiple years of experience an Integral System Simulator (ISS) has been developed and put into operation. This ISS is successfully used for the practical training of the controllers [6]. The ISS is designed for theoretical training, practical training, and conversion training of the controllers of aerodrome and area ATC systems with different level of automatization. The ISS allows for a high-quality training of the ATC controllers in accordance with the stipulated operating procedures at the workstations of the ATC units: on route and off the track, in aerodrome areas and aerodrome terminal areas. It provides for building up, improving, and mastering of professionally important qualities, skills, and abilities, which serve for higher decision-making speed in normal, conflict, emergency, and in-flight emergency situations.

The ISS solves the tasks of the training of the ATC controller of area and terminal zones, and adjusts itself to the relevant zones and sectors (area, approach, circuit, landing/runway, taxiing). The ISS is based on the modular design principle, which implements the open information system architecture, that allows for:

- structural flexibility of each functional module (area, approach, circuit, landing/runway, taxiing);
- depending on the Customer's requests, expanding of the system for an optional number of functional modules;
- invariant nature of the configuration of functional modules (of one type and up to an optional combination of the volume of the requested number of modules).

The ISS is configured with the number of the universal simulation modules (USM) (defined by the Customer) realized on the distributed computer aids structure joined into a single unit by means of LAN.

The USM includes:

- the automated workstation of the controller–trainee (AWS-C);
- the automated workstation of the operator pilot/instructor (AWS-OP).

AWS-Cs can be included into the ISS in different design options: starting from the option with low-automation ATC system functions and up to the options with high-automation ATC system functions, which comply with ICAO and EUROCONTROL requirements. At that, the AWS-C of the area center, approach, and circuit centers can include either one section of the radar surveillance controller or two sections: the radar surveillance controller and procedure monitoring controller (planning controller).

The AWS-OPs include the simulation aids, air situation data processing and display aids; radiocommunication and intercom.

The ISS "Syntez-TC" ("Синтез-ТЦ") allows for:

- airspace simulation up to 2500×2500 km and up to the height of 25 km;
- traffic simulation of an optional number of manned and unmanned aircraft;
- accomplishment of up to 200 flight plans within one exercise, when the duration of the exercise is up to 120 min;
- air situation simulation in real-time and fast-time scale both within the visibility range of the simulated radar aids (taking into account the terrain) and outside of the visibility range, with the possibility to simulate conflict, emergency, and in-flight emergency situations;
- current planned data simulation required for subsequent passing from the operator pilot to the controller–trainee;
- radar data simulation taking into account the visibility ranges of the simulated radar aids, and special signals of secondary surveillance radars (SSRs);
- direction finding equipment data simulation in order to display the bearing at the combined air situation display (ASD) of the controller–trainee;

- visual controlled ground situation simulation including the runway, taxiways, aircraft parking places, facilities, all lighting equipment, as well as moving objects (aircraft, tow cars, etc.) taking into account the position of the viewer;
- aerodrome lighting conditions change simulation in accordance with the input time of the day;
- runway emergencies simulation (aircraft collision, aircraft fire, complex ornithological conditions, etc.);
- complication of the aerodrome situation surveillance through simulation of fog with adjustable intensity and color, changing of the time of the day, presence of weather hazards;
- automatic registering of the air situation parameters and trainees' actions based on the main flight safety provision criteria within the ATC systems;
- possibility for instructors to intervene into the exercise (addition and removal of restrictions, introduction of new targets, change of the meteorological conditions, exercise pause and playback, etc.) in coordination with the training supervisor;
- meteorological conditions simulation including weather hazards and actual weather on the runway;
- radiocommunication and intercom simulation with the possibility to connect personal headsets;
- automated assessment of the controller–trainees activities parameters;
- record with subsequent simultaneous playback of the radar and voice data;
- record and pause of the exercise with information storage and subsequent continuation of the exercise;
- software control for the collaborating ATC areas subject to taking over/passing of the aircraft control during entering/exiting the area of responsibility.

3.2.2 Controller's STD

The modular integral controller's STD allows for simulation of the operation of ATC AS of any standard or type. The STD software allows for simulation of the modern display systems of the ATC AS systems in accordance with ICAO and EUROCONTROL recommendations. The STD is configured to include the workstations of the controllers and workstations of operator pilots. The work of the controller–trainees is monitored by the instructors of the training center. The controller's STD allows for solving of the following main tasks—aircraft and ground vehicles movement simulation with the possibility to provide control from the parking place and take-off up to the exit from the current ATC area or even further (up to the landing and parking). The maximum number of objects is up to 500 per 1 module. The automated buildup of the ATC area with the description of up to 1000 radar beacons, 500 routes, 32 areas of responsibility, indefinite number of samples of user graphics, up to 10 layers of terrain data per each declination, setting of several radars, description of the terrain for each preset radar—automated maintenance and editing of up to 300 aircraft types—automated processing and editing of the weather databases with optional changes of the time and with the maximum intensity of up to 300 wind layers and 100 thunderstorm cells—ASD at the scale from 500 meters up to 5000 km—switching of scales (up to the display of the taxiing area) at any workstation—simulation of the taxiing–runway (TOWER) with the display of the runways, taxiways, parking places, facilities, and aircraft in 3D mode at the ASD—image scaling up to 4:1 order—smooth scaling with optional move—the multi-windows radar data display function is implemented. The function of operation of adjacent ATC areas with the radar data window is implemented. The operation of the modules of adjacent ATC areas in a complex mode is implemented with the operation simulation in different ATC areas at the same time. The aircraft database with graphical processing of the movements parameters, and visual display of the aircraft—the movement parameters include speeds within the taxiing/runway areas, accelerator of the speed change. Different configurations of the aircraft work windows—change of the aircraft movement parameters with the help of the keyboard or with the mouse pointing device—at the discretion of the user. There are four alternative methods for the aircraft control commands input: through the context menu, with the help of the short cuts, manual input of the parameters in the additional request window, or through the aircraft window. Input of standard commands for the aircraft movement parameters control—possibility to change the routes with the help of the prepared in advance standard routes within the ATC areas—standardized display of the radar beacons following the ICAO and EUROCONTROL standards—preparation of standard routes up to the taxiing/runway area—switching of the display by magnetic or true declination—layer-by-layer display of data (closed areas, restricted areas, etc.)—possibility to display the number of the radar beacons and their name—possibility to switch on/off and choose the content of the flight following log—possibility to restrict the display of the logs by layers—possibility to choose the form of the aircraft symbol (four options) with a marking on clearance to fly within the airspace with defined separation minima—synchronization of aircraft displacement with the scanning beam of the radar—automatic calculation of approaches to landing referenced to a specific runway and approach direction—entering the holding area at any point of the track or above the radar beacon—possibility for the instructor to restrict functional capabilities of the controller's display equipment—simulation of the direction finder grid at the point of time of aircraft communication at the controller's display—simulation of meteorological conditions within the ATC areas that are simulated with the STD subject to the influence on the aircraft movement—simulation of thunderstorm cells with different flash intensity (not less than 8 grades)—possibility to display the flight planning data for the workstation at the controller's display with differentiation between arriving and departing aircraft—possibility to print out the flight plans—possibility to display the exercise at the workstation of the instructor of the flight plan. The software package includes the flight plan editing and viewing (in ICAO format) software. Viewing and editing of the flight plans is also supported by the pilot's and controller's software. Automatic transfer of all changes and additions to all PCs connected into the STD network. Automatic re-calculation of all databases after input of additional radar beacons regardless of the place of their registry. Choice of the display option for the range rings starting from one and up to 100 depending of the scale—possibility to display 5-, 10-, 30-grade azimuth lines—possibility to display the geographic coordinate grid—possibility to change the display of the tunnel routes (axil lines, tunnels of different width)—change of thickness of the track depiction lines—software that allows changing color, brightness, and contrast of the image—possibility to display geographic maps (up to ten layers). The terrain, which is taken into account during the display of radar data, is processed (visibility of aircraft symbols at the controller's display and altitude alerts). Referencing of the radar beacons (track method of the ATC areas buildup)—possibility to display the workstations' areas of responsibility and the radar coverage at the screens—automatic recording of the exercise with the possibility for a playback of the whole training with the synchronized playing of the sound sessions, as well as multiple replay of the exercise from any point of time for subsequent training—possibility to create the radar beacon and air tunnel database for the whole world—possibility to provide separation in accordance with the RVSM standards—software simulation of the radio communications and controllers' intercom without any additional equipment—just through the computer network capabilities—software or manual switching of the radio communication channels—possibility of quick re-configuration of the STD depending on its tasks and availability of instructors and trainees—possibility of quick call for assistance (reference information on the STD) at any time and at any place—possibility to create new areas, change, and supplement the databases by the user without approaching the developer—possibility to switch between the Russian and English interface of the STD without exiting the software—aircraft movement parameters input in different measuring systems—network view and preparation of the archive of recorded training sessions through any module of the STD and simultaneous playback of the exercise log at chosen computers of the STD—quick change of the STD operation parameters by the Customer (separation rules, conflicts, aircraft technical parameters, ATC areas composition, user schedule). Free configuration of the STD allows for establishing of either one connected subsystem of several independent simulation systems.

3.3 ATC STDs' Classification

3.3.1 ATC Controllers' STD Classification

Analysis of the methodology of the ATC controllers' practical training arrangement allows outlining the following main classes of the ATC controllers' STDs.

1. STDs of high quality (of high precision) for group training. These STDs usually have precise replicas of the workstations of the respective ATC AS controllers including all equipment and software required for presenting full list of tasks within the control sectors or at the tower and their environment. For the work-

stations of the runway and taxiing controllers, it includes the model of the visual aerodrome situation (view from the tower).

2. Individual training STD. These STDs simulate for the trainee the most important characteristics (features) of the real situation and reconstruct operational conditions, which allow for practicing of tasks in real time.

3. Procedure STD. These STDs are the training devices (personal Electronic Computing Machine (ECM)), which allows the trainee to practice some operational (work) functions separately from other functions, which are not represented over there, although they are always connected to the first ones during the real operation.

4. Other training devices are the computer-aided software, which provide the trainee with some operational (work) functions under unrealistic operation of the operating devices. As a rule, these are generating computers or workstations (connected into a network or independent) which are designed for the use of one trainee or a small group. At that the commonly used equipment (hardware and software) is used rather than equipment, which is highly modified for the ATC needs.

3.3.2 Methodology of Establishing the Training Process for the Controllers

Establishment of the training process, which provides (at first) accumulation of knowledge (abilities), and then certain practicing of specific tasks and after that simulation (training) and final stage of the training—practicing of the whole set of functional tasks at the ISS in group mode, allows for optimization of the training process efficiency.

Expansion of the requirements toward the resources during the establishment of the training is resolved in the following manner: The resources are provided as and when necessary; accumulation of abilities (knowledge) is realized at a generating device, specific tasks are practiced at the procedure STD, and (in the course of the tasks, strategies, and judgments' integration) the training is shifted to the realistic simulation of ATC processes at the STD of high precision. In Table 3.1, a methodology of use of different training aids during realization of the task of the controllers' professional training is shown.

3.3.3 Characteristics of the Training Aids Used Within the EUROCONTROL States

The study of the current practice is required to explain the concepts and directions of the development of the software and hardware aids of the ATC controllers' STDs. The all-round study of the implementation options subject to the country allows

Table 3.1 Use of equipment of simulators

Tasks	Technical aids			
	Training devices	Procedure STDs	Individual STDs	Integral STDs of high precision
Accumulation of knowledge (skills)	Best use	Not required	Not required	Not required
Specific tasks practicing	Not enough	Best use	Best use	Not required
Individual simulation	Not enough	Not enough	Best use	Best use
Command simulation	Not enough	Not enough	Best use	Best use
Group simulation	Not enough	Not enough	Best use	Best use

highlighting the most advanced trends in the STDs development and use. In the summary of Table 3.2, you can see the countries that are part of the EUROCONTROL and that have achieved significant success in the use of different technical aids in order to provide for the relevant controllers training.

In what follows, in Tables 3.3, 3.4, 3.5, 3.6, 3.7, 3.8 and 3.9, you can find the characteristics of different technical aids in more detail and their scope of application subject to solving the tasks of the controllers training within the main training centers of the EUROCONTROL states.

The comparative study of the ATC controllers' simulators that are developed and implemented by the Russian manufacturers and by the leading foreign companies allows making the following conclusions.

For the controllers' training, a fairly broad number of technical training aids is developed and used; these are: universal sets of technical aids, procedure STDs, STDs for individual training, and integral STDs of high precision.

For different training stages: theoretical training, functions practicing, practical training, conversion training—the economic efficiency of the use of different technical aids for controllers training is defined.

Section Review

1. Tasks solved by the ATC controllers' STDs and their influence on the flight safety-level provision.
2. Trends in the development of the ATC controllers' STDs hardware/software aids.
3. Comparative study of the samples of the controllers training aids.
4. Classification of the training aids.
5. Main characteristics of the ATC controllers' STDs.

Table 3.2 Contingent of EUROCONTROL simulators

Tasks	Technical aids			
	Training devices	Procedure STDs	Individual STDs	Integral STDs of high precision
1	2	3	4	5
Accumulation of knowledge (skills)	Denmark EUROCONTROL Germany Italy The Netherlands	EUROCONTROL France Italy Spain	Denmark Ireland	France
Controlled accumulation of knowledge (skills)	Spain	Germany	Ireland	
Specific tasks practicing		Germany Great Britain Belgium	France Ireland Czech Republic	Not required
Controlled specific tasks practicing		Germany		
Individual simulation		Germany	EUROCONTROL France Germany Great Britain Ireland Denmark Czech Republic	France Italy Great Britain
Command simulation			Belgium Czech Republic EUROCONTROL Spain Ireland Italy	Italy Spain The Netherlands
Group simulation			EUROCONTROL Spain Ireland Czech Republic Belgium	France Germany Italy Spain Great Britain
Digital training	EUROCONTROL France Germany Spain Great Britain			

Table 3.3 Institute of Air Navigation Services (Luxembourg)

Tasks	Technical aids			
	Training devices	Procedure STDs	Individual STDs	Integral STDs of high precision
1	2	3	4	5
Accumulation of knowledge (skills)	Computer classroom: WS: 20 St/WS: 1 OP/WS: 0 St/Instr: 12	First Twr 2d: WS: 2 St/WS: 1 OP/WS: 1 St/Instr: 1		
Controlled accumulation of knowledge (skills)				
Specific tasks practicing				
Controlled specific tasks practicing				
Individual simulation			First: 6 WS: 6 St/WS: 1 OP/WS: 1 St/Instr: 1	
Command simulation			First: 6 WS: 3 St/WS: 2 OP/WS: 1 St/Instr: 1	
Group simulation			First: 6 WS: 6 St/WS: 1 OP/WS: 1 St/Instr: 1	
Controlled simulation				
Digital training	Computer classroom: WS: 20 St/WS: 1 OP/WS: 0 St/Instr: 12			

Note First line—the name of the STD and the number of simultaneously trained students
WS—the number of the workstations
St/WS—the number of the students per one workstation
OP/WS—the number of the operator pilots per one workstation
St/Instr—the number of the students per one instructor

Table 3.4 Ecole Nationale de l'Aviation Civil (ENAC) (Toulouse, France)

Tasks	Technical aids			
	Training devices	Procedure STDs	Individual STDs	Integral STDs of high precision
1	2	3	4	5
Accumulation of knowledge (skills)				Aer./Sim.: 8 WS: 4 St/WS: 2 OP/WS: 1 St/Instr: 2
Controlled accumulation of knowledge (skills)		PTT: 16 WS: 16 St/WS: 1 OP/WS: 0 St/Instr: 8		
Specific tasks practicing			Scanor: 16 WS: 8 St/WS: 2 OP/WS: 1 St/Instr: 2 — Scanrad: 16 WS: 8 St/WS: 2 OP/WS: 1 St/Instr: 2	
Individual simulation			Scanor: 8 WS: 8 St/WS: 1 OP/WS: 1 St/Instr: 2	Aer./Sim.: 8 WS: 4 St/WS: 2 OP/WS: 1 St/Instr: 2 — Electra: WS: 16 St/WS: 2 OP/WS: 1 St/Instr: 2
Group simulation				Electra: 32 WS: 16 Ct/WS: 2 OP/WS: 3 St/Instr: 2

(continued)

Table 3.4 (continued)

Tasks	Technical aids			
	Training devices	Procedure STDs	Individual STDs	Integral STDs of high precision
Digital training	Computer classroom: WS: 32 C_T/WS: 1 OP/WS: 0 St/Instr: 32			

Note First line—the name of the STD and the number of simultaneously trained students
WS—the number of the workstations
St/WS—the number of the students per one workstation
OP/WS—the number of the operator pilots per one workstation
St/Instr—the number of the students per one instructor

Table 3.5 DFS academy (Langen, Germany)

Tasks	Technical aids			
	Training devices	Procedure STDs	Individual STDs	Integral STDs of high precision
1	2	3	4	5
Accumulation of knowledge (skills)	Computer classroom: WS: 16 St/WS: 1 OP/WS: 0 St/Instr: 16			
Controlled accumulation of knowledge (skills)		Basim: 16 WS: 16 St/WS: 1 OP/WS: 0 St/Instr: 1		
Specific tasks practicing		Basim: 16 WS: 16 St/WS: 1 OP/WS: 0 St/Instr: 1		
Controlled specific tasks practicing		Basim: 16 WS: 16 St/WS: 1 OP/WS: 0 St/Instr: 1		
Individual simulation		Basim: 16 WS: 16 St/WS: 1 OP/WS: 0 St/Instr: 1	Basim: 16 WS: 16 St/WS: 1 OP/WS: 0 St/Instr: 1	

(continued)

Table 3.5 (continued)

Tasks	Technical aids			
	Training devices	Procedure STDs	Individual STDs	Integral STDs of high precision
Group simulation				Asim: 64 WS: 32 St/WS: 2 OP/WS: 4 St/Instr: 2 — Tosim: 9 WS: 3 St/WS: 3 OP/WS: 4 St/Instr: 3
Digital training	WS: 32 OP/WS: 0 Instr: 0			

Note First line—the name of the STD and the number of simultaneously trained students
WS—the number of the workstations
St/WS—the number of the students per one workstation
OP/WS—the number of the operator pilots per one workstation
St/Instr—the number of the students per one instructor

Table 3.6 ENAV (Rome, Italy)

Tasks	Technical aids			
	Training devices	Procedure STDs	Individual STDs	Integral STDs of high precision
1	2	3	4	5
Accumulation of knowledge (skills)	Computer classroom: WS: 10 St/WS: 1 OP/WS: 0 St/Instr: 10	PcTwr pre-trainer: WS: 10 St/WS: 1 OP/WS: 0 St/Instr: 0		
Controlled accumulation of knowledge (skills)				

(continued)

Table 3.6 (continued)

Tasks	Technical aids			
	Training devices	Procedure STDs	Individual STDs	Integral STDs of high precision
Specific tasks practicing		PcTwr trainer: WS: 10 St/WS: 1 OP/WS: 0 St/Instr: 0		
Controlled specific tasks practicing				
Individual simulation				Alenia-Atres: WS: 4 St/WS: 1 OP/WS: 1 St/Instr: 1
Command simulation			Twr 2d visual: WS: 4 St/WS: 2 OP/WS: 2 St/Instr: 2 ——— Sim App: 8 WS: 4 St/WS: 2 OP/WS: 2 St/Instr: 2	Alenia-Atres: WS: 4 St/WS: 2 OP/WS: 1 St/Instr: 2
Group simulation				Alenia-Atres: WS: 4 St/WS: 2 OP/WS: 1 St/Instr: 2
Controlled simulation				
Digital training				

Note First line—the name of the STD and the number of simultaneously trained students
WS—the number of the workstations
St/WS—the number of the students per one workstation
OP/WS—the number of the operator pilots per one workstation
St/Instr—the number of the students per one instructor

Table 3.7 Nordic ATS academy (Malmö, Sweden)

Tasks	Technical aids			
	Training devices	Procedure STDs	Individual STDs	Integral STDs of high precision
1	2	3	4	5
Accumulation of knowledge (skills)	Computer classroom: WS: 15 St/WS: 1 OP/WS: 0 St/Instr: 15	Twr: WS: 1(4) St/WS: 1(4) OP/WS: 3(12) St/Instr: 1(4)	Pro T: WS: 1(6) St/WS: 1(6) OP/WS: 1(6) St/Instr: 1(6)	
Controlled accumulation of knowledge (skills)				
Specific tasks practicing				
Individual simulation			BaRT/BERT: WS: 23 St/WS: 1 OP/WS: 1–2 St/Instr: 1	SMART: WS: 14 St/WS: 1 OP/WS: 1–2 St/Instr: 1
Command simulation		Twr: WS: 1(4) St/WS: 1 OP/WS: 3 St/Instr: 1	BERT: WS: 3 St/WS: 2 OP/WS: 1 St/Instr: 1	SMART: WS: 8 St/WS: 1 OP/WS: 2 St/Instr: 1
Group simulation			BERT: 1–2 WS: 16 St/WS: 1 OP/WS: 2 St/Instr: 1	SMART: WS: 8 St/WS: 1 OP/WS: 2 St/Instr: 1
Controlled simulation				
Digital training	Computer classroom: 15 WS: 14 St/WS: 1 OP/WS: 16 St/Instr: 12	Computer classrooms: Twr: 4 ProT: 6 Bart: 7 Rad/14pil Bert2:8 rad/16pil Smart: 28 works		

Note First line—the name of the STD and the number of simultaneously trained students

WS—the number of the workstations

St/WS—the number of the students per one workstation

OP/WS—the number of the operator pilots per one workstation

St/Instr—the number of the students per one instructor

Table 3.8 IVAO (Zurich, Switzerland)

Tasks	Technical aids			
	Training devices	Procedure STDs	Individual STDs	Integral STDs of high precision
1	2	3	4	5
Accumulation of knowledge (skills)	Computer classroom: WS: 12 St/WS: 1 OP/WS: 0 St/Instr: 12			
Controlled accumulation of knowledge (skills)		BASIM: 6 WS: 6 St/WS: 1 OP/WS: 2 St/Instr: 1	BASIM: 6 WS: 6 St/WS: 1 OP/WS: 2 St/Instr: 1	
Specific tasks practicing		BASIM: 6 WS: 6 St/WS: 1 OP/WS: 2 St/Instr: 1	BASIM: 6 WS: 6 St/WS: 1 OP/WS: 2 St/Instr: 1	
Controlled specific tasks practicing		BASIM: 6 WS: 6 St/WS: 1 OP/WS: 2 St/Instr: 1	BASIM: 6 WS: 6 St/WS: 1 OP/WS: 2 St/Instr: 1	
Individual simulation		BASIM: 6 WS: 6 St/WS: 1 OP/WS: 2 St/Instr: 1	BASIM: 6 WS: 6 St/WS: 1 OP/WS: 2 St/Instr: 1	INTRAS ZH: 12 WS: 6 St/WS: 2 OP/WS: 2.5 St/Instr: 2 INTRAS GE: 8 WS: 4 St/WS: 2 OP/WS: 2.5 St/Instr: 2

(continued)

Table 3.8 (continued)

Tasks	Technical aids			
	Training devices	Procedure STDs	Individual STDs	Integral STDs of high precision
Command simulation		BASIM: 12 WS: 12 St/WS: 2 OP/WS: 2 St/Instr: 2	BASIM: 12 WS: 12 St/WS: 2 OP/WS: 2 St/Instr: 2	INTRAS ZH: 12 WS: 12 St/WS: 1 OP/WS: 1.25 St/Instr: 2 INTRAS GE: 8 WS: 8 St/WS: 1 OP/WS: 1.25 St/Instr: 2 TOSIM ZH: 3 WS: 3 St/WS: 1 OP/WS: 1.33 St/Instr: 3
Group simulation				INTRAS ZH: 12 WS: 12 St/WS: 1 OP/WS: 1.25 St/Instr: 2 INTRAS GE: 8 WS: 8 St/WS: 1 OP/WS: 1.25 St/Instr: 2

Note First line—the name of the STD and the number of simultaneously trained students
WS—the number of the workstations
St/WS—the number of the students per one workstation
OP/WS—the number of the operator pilots per one workstation
St/Instr—the number of the students per one instructor

Table 3.9 Civil school of air traffic control (Great Britain, Bournemouth)

Tasks	Technical aids			
	Training devices	Procedure STDs	Individual STDs	Integral STDs of high precision
1	2	3	4	5
Accumulation of knowledge (skills)				
Controlled accumulation of knowledge (skills)				
Specific tasks practicing		SIMTAC: 16 WS: 16 St/WS: 1 OP/WS: 2 St/Instr: 1		
Individual simulation			SIMTAC: 16 WS: 16 St/WS: 1 OP/WS: 2 St/Instr: 1 Skywatch: 9 WS: 9 St/WS: 1 OP/WS: 3 St/Instr: 1	First: 20 WS: 20 St/WS: 1 OP/WS: 2 St/Instr: 1
Command simulation			Skywatch: 6 WS: 3 St/WS: 2 OP/WS: 3 St/Instr: 1	
Group simulation				First: 20 WS: 20 St/WS: 1 OP/WS: 2 St/Instr: 1
Digital training	Computer classroom: WS: 6 St/WS: 1 OP/WS: 0 St/Instr: 6			

Note First line—the name of the STD and the number of simultaneously trained students

WS—the number of the workstations

St/WS—the number of the students per one workstation

OP/WS—the number of the operator pilots per one workstation

St/Instr—the number of the students per one instructor

References

1. Filin AD, Velmisov IA, Neustroev NN (2010) Training and modeling systems in combat training of Air Forces. Science and Production, Moscow, p 30
2. Regulation on the classification of CA specialists. Air Transport, Moscow 2001
3. Dundukov VP, Shcherbakov EK (1983) Classification of functional tasks of controller simulators. Air Transport, Moscow, p 232
4. Filin AD, Suleimanov RN (2008) Studying ways to create technical means to ensure flight safety, to assess the level of combat training in Air Forces subunits and units during trainings, exercises and live firing. Research report (stage I). SPb., VNIIRA, p 206/44
5. Filin AD, Suleimanov RN et al (2009) Studying ways to create technical means to ensure flight safety, to assess the level of combat training in Air Forces subunits and units during trainings, exercises and live firing. Research report (stage II). SPb., VNIIRA, p 256/64
6. Filin AD, Suleimanov RN et al (2011) Integrated system simulator of the ATC LSU. Explanatory note for the technical project. SPb., VNIIRA, p 274

Chapter 4
Technical Requirements to the ATC Automation System Simulators for Controllers

Designation requirements The ISS shall provide for the simulation of the main technological processes of the airspace use planning and air traffic control, which are realized at the workstation of the area control center (ACC), aerodrome centers and aerodrome control towers (ACT) of specified aerodromes. The ISS shall provide for the possibility to train the controllers of the following ACCs and ATM aerodrome centers: circuit control unit (CCunit), approach control, Local ATC Unit, area center (AC) of the Unified ATM System (on route and off-the-track sectors), as well as the aerodrome (Area) flight operations director (AeroFOD, AreaFOD). The ISS shall provide the possibility to train the controllers of the following ACTs: taxiing supervisory unit (TSU), start control unit (SCU), landing control tower (LCT). The ISS shall provide the possibility to train the flight planning controllers of the aerodrome centers and area centers [1, 2].

Tactical requirements Within the ISS, the following main tactical requirements shall be implemented:

- simulation of movements of different aircraft types following the commands of the controllers both on the ground and in the air in accordance with their technical and aerodynamic characteristics, taking into account the weather conditions, and following the stipulated flight plans;
- simulation of the functioning of the flight support radiotechnical aids;
- simulation of the flight support planning system;
- simulation of different changes of the airspace structure within the aerodrome area;
- simulation of the influence of different weather conditions (rain, snow, snowstorm, fog, ceiling, runway visibility, wind direction and speed, etc.);
- simulation of different times of the day (day, night, twilight) and of the year (summer–winter);
- possibility to train the controllers either independently by the control units (individually) or in an integrated manner: simultaneously in a connected group of the controllers of the stipulated units;
- simulation and display of the weather information;
- simulation and display of information on the current and forecasted air situation;

© Springer Nature Singapore Pte Ltd. 2020
Bestugin A.R. et al., *Air Traffic Control Automated Systems*,
Springer Aerospace Technology, https://doi.org/10.1007/978-981-13-9386-0_4

- simulation and display of the information restrictions of the airspace;
- simulation and display of the simulated special signals coming from the aircraft;
- simulation and display of the warnings of aircraft entering the hazardous weather conditions, closed or dangerous zones, aircraft deviation from the flight track, aircraft descent below the safe height;
- simulation and display of the formalized messages (by ICAO, EUROCONTROL standards) and of the special ATS messages;
- simulation of the voice information ("controller-aircraft," "controller-controller," "intercom");
- simulation of the movements of the automobile vehicles and special-purpose vehicles on the ramp, taxiways, main taxiways, and runway;
- simulation and display of the ATIS information;
- simulation of the aerodrome lighting equipment;
- simulation of the enclosing terrain with detailed relief, natural and artificial reference points, which are seen from the observation point (mountain, hill, mast, pipe, panoramic view of the populated area, power transmission lines, etc.);
- simulation of different MTCD criteria;
- simulation of conflicts and in-flight emergencies.

Within the ISS, the requirement for the possibility to solve the functional tasks of the ACC controllers shall be implemented:

- clearing for engine start, line up and hold subject to the flight plan, air and ground situation;
- safe aircraft separation for takeoff and landing;
- clearing for aircraft takeoff and landing subject to the weather conditions, airfield conditions, operability of aerodrome radiotechnical and lighting equipment;
- clearing for ground activities conducted by the aerodrome services within some parts of the airfield and monitoring of timely and complete vacancy of these parts by vehicles and people;
- monitoring the absence of obstacles on the runway, parts of the airfield and aerodrome when clearing the aircraft for taxiing, takeoff and landing, as well as preventing emergency situations due to inadvertent or unauthorized movements of different objects (aircraft, vehicles, people, animals, etc.);
- monitoring the aircraft external view during visual monitoring in order to practice the detection of malfunctions on the aircraft (fire, non-extension, partial extension, or non-retraction of the landing gear, etc.);
- practicing of actions in the event of weather conditions changes, which influence aircraft takeoff and landing clearances, as well as of actions to counteract hazardous (for aviation) weather conditions;
- practicing of actions in the event of emergencies and in-flight emergencies with aircraft or ground objects;
- request for display (following the command of the controller) of the airspace use plans, which are stored in the ISS, their editing, canceling, saving, copying, and printing;

- correcting of the planned information on the follow-me vehicle with the help of the surveillance system (simulation);
- correcting of the planned information on the manually input messages;
- forming of the planned lists based on the data from the current exercises, and their provision to the controllers;
- display of the meteorological information: at the aerodromes and within the area of responsibility;
- controllers' shift handover procedure practicing;
- forming of the time sequence for arriving and departing aircraft;
- landing time assignment following the adjacent sector entrance clearance;
- coordination of the conditions of entering the adjacent sector;
- defining the time windows free for takeoff and landing planning;
- defining the time windows restricting the landing operations due to the turbulence and separation minima;
- alerting on violation of the acceptable time separation between the aircraft performing landing/takeoff during the operational planning;
- display of the recommended landing time by the controllers' stations;
- defining the duration of the required aircraft delay during takeoff/landing;
- defining the duration of the required aircraft delay by its current coordinates and weather conditions;
- allocation of the duration of the required arriving aircraft delay by the workstations;
- defining the acceptable separation (distance) between the aircraft at landing;
- defining the recommended aircraft landing time;
- gathering information on aircraft flight preparation accomplishment;
- defining the estimated time of the aircraft transfer to the ATS control;
- defining the recommended take-off time;
- defining the time windows free for take-off planning;
- defining the time windows restricting the takeoff both due to the turbulence and in order to provide the vertical and horizontal separation minima;
- defining the recommended take-off time by the controllers' stations;
- coordinating the times of takeoff with the times of landing;
- allocation of the duration of the required departing aircraft delay by the workstations.

Within the ISS, the requirement for the possibility to solve the functional tasks of the TWR, ACC, and Local ATC units' controllers shall be implemented:

- based on the surveillance information on the current and forecasted air situation, detection of potential conflicts and warning of the controllers on potential near misses;
- based on the surveillance information, detection and warning of the controllers on reaching of the limit values;
- based on the surveillance information on the current and forecasted air situation, detection and warning of the controllers on the aircraft descent below the minimum safe height;

- detection of the "medium-term" aircraft conflicts, which is provided based on the calculation of the space-temporal aircraft trajectories (at the second stage of the current planning), and display of the respective warning to the controller together with additional information on the conflict (by means of special windows for detailed analysis of the potential conflicts);
- monitoring of the trajectory maintaining and alerting (MONA);
- medium-term conflict detection (MTCD);
- maintaining of the assigned flight level and warning of the controller on deviations;
- detection of the aircraft deviation from the planned flight track (in the horizontal plane) and warning of the controller;
- control of arriving and departing aircraft with the definition of the sequence of approach and departure for the aerodromes within the Moscow airspace;
- medium-term conflict resolution advisory (CORA);
- trajectory forecasting with the transition from standard route to alternative and back to the standard;
- calculation and registration of the minimum separation between the aircraft, which was followed by the potential near-miss detection;
- based on the surveillance information (simulated) on the current and forecasted air situation, detection and warning of the controllers on the possibility for the aircraft to enter the zone with the restricted airspace use;
- based on the surveillance information (simulated) on the current and forecasted air situation, detection and warning of the controllers on the possibility for the aircraft to enter hazardous weather conditions.

Within the ISS, the requirement for the possibility to solve the functional tasks of the TWR and ACC airspace use planning controllers shall be implemented:

- flight plans and ATC messages data handling;
- weather data handling;
- air navigation and airspace use restrictions data handling;
- reference and support data handling;
- daily flight plan formation and maintaining;
- outgoing messages via Civil Aviation Terrestrial Data Transmission and Telegraph Communication (TDTTC) network formation and simulation;
- simulation of the TDTTC incoming messages flow;
- output forms formation;
- messages and output forms printing;
- monitoring of availability of flight permission;
- calculation of the forecasted traffic within the ATC sectors and area;
- calculation and presentation in visual form of the planned flight routes;
- preparation, analysis, and maintaining of the daily flight plan;
- preparation and submission of standard messages on the airspace use in accordance with the messages table;
- monitoring and analysis of the incoming messages on the airspace use;
- monitoring and analysis of the weather situation at his/her and alternate aerodromes, as well as along the tracks and within the flight areas;

- monitoring and analysis of the air navigation information on the aerodrome, radiotechnical aids and air routes condition;
- cooperation with the adjacent control units and airport services.

Technical Requirements Within the ISS, the following main technical requirements shall be implemented. Simulation of the aircraft radar tracking by the primary radar, secondary radar, primary + secondary radar shall be provided within the ground speeds range from 0 to 3500 km/h, as well as the aircraft movement extrapolation. The maximum number of the simultaneously simulated targets at the air situation display—no limits.

The maximum number of the flight plans shall be: standard (repeated)—in accordance with the exercise scenario; passive—no limits, active—in accordance with the exercise scenario.

The maximum number of the generated control sectors: within the aerodrome area—in accordance with the ISS structure; along the routes—in accordance with the ISS structure and the exercise scenario; off track—in accordance with the ISS structure and the exercise scenario; at Local ATC Unit—in accordance with the ISS structure and the exercise scenario.

The response time for the panel operation shall not exceed 0.5 s.

Display of the objects' movement parameters shall be provided: in the main mode—in accordance with the SI system; in supplement mode—in accordance with the imperial system (feet, knots, etc.).

The ASD's interface of the radar control controller shall be unified with the AWS of the controllers of the simulated ATC AS to the fullest extent possible. The basis for the simulation of the air situation shall be preliminary prepared aircraft flight plan sets, which form the exercises, during the operational control of the aircraft movements following the ATC controller commands. The simulation of the airspace shall be provided within the range of not less than 2300 × 2500 km in distance and 25 km in height with the simulation of the air situation in real-time and fast-time scale.

In order to control the aircraft through the ISS, the following functions shall be considered: control transfer between the operator pilots; setting of the altitude (flight level) both in meters and in feet; setting of the vertical climb/descent rate; setting of the altitude (flight level) to the specified point; setting of the lateral displacement; setting of the heading; setting of the turn with a specific angle; setting of the bank angle during the turn; setting of the turn direction; setting of the new squawk; simulation of the squawk modes; flight to the specified point; flight toward the holding area and operation within this area; change of the flight track; aircraft operation by the trajectories.

Within the ISS, the simulation of the weather data shall be provided: actual weather at the runway; wind parameters by the heights; hazardous weather conditions; weather conditions changes at the aerodrome. Within the ISS, the software control for the cooperating ATC areas shall be provided in terms of notification, activation, and acceptance/transfer of the aircraft control during entering/exiting (from) the area of the responsibility.

Within the ISS, the simulation of the aircraft radar tracking within the speed range from 0 to 3000 km/h and monitoring of the transport vehicles ground movements within the speed range from 0 to 300 km/h shall be provided. Within the ISS, the simulation of the following emergencies and in-flight emergencies shall be provided: aircraft fire; engine shutdown; cabin depressurization; entering the hazardous weather conditions (thunderstorm cells, turbulence, icing, etc.); radiocommunication loss; attack on the flight crew; loss of spatial orientation.

During the simulation of the emergencies and in-flight emergencies within the exercise scenario and instructor's (exercise trainer's) panel operations, the relevant flight crew actions and reports shall be introduced. During the simulation of the emergencies within the ISS, the following scenarios of emergencies and in-flight emergencies shall be realized: aircraft landing with landing gear up; tire burst; landing gear break below during takeoff or landing; aircraft in-flight fire and smoke; aircraft collision with the ground or ground obstacles; aircraft excursion from the airfields during takeoff, landing and taxiing; aircraft collision on the ground, during takeoff or landing; aircraft collision with the ground transport; possibility to simulate the movements of the emergency services in all the above-mentioned cases.

Training requirements Within the ISS, the following training requirements shall be implemented.

The ISS shall provide for recording of the whole exercise during the training, including the dynamic air situation, planning data, weather data, voice data.

The recording aids shall provide for registry of the information given in the scenario and created within the course of the exercise up to the moment of its completion; search, view, and playback in real-time and in a fast-time scale (by user's choice) of the recorded information starting from any point of time (by user's choice); simultaneous playback of the recorded information in real time.

The following information is subject to recording: air situation data, registered data on the STD technical condition, violation of the standards and rules of the flight management or ATS (potential conflicts, conflicts due to clearances to use the occupied airfield or flight through hazardous or closed areas, safe heights violation, etc.), trainee data, controllers' mistakes logbook and any other information required for the subsequent analysis of the training results. The following shall be archived: information databases; created ATS areas; prepared exercise scenarios; exercises performed in each module; training results; lists of instructors and trainees.

Within the ISS, the possibility for the following shall be provided:

– realization of an unlimited number of the flight plans within one exercise in accordance with the tactical target of the exercise considering the required duration of the exercise;
– pausing the exercise with preserving the information on the screens with subsequent continuation or restarting of the exercise from the beginning;
– software control for the cooperating ATC areas in terms of notifications, activation, and acceptance/transfer of the aircraft control during entering/exiting (from) the area of the responsibility;

- input of the aircraft control commands;
- input of the commands, which facilitate or complicate the air situation (weather conditions change, introduction of additional aircraft, introduction of the emergencies).

During the exercise preparation, the possibility for the following shall be provided:

- user-friendly data input and editing of these data in the "menu" mode;
- exercise formation in terms of the area (zone) structure (trajectory points, holding areas, typical references, compulsory reporting points, etc.), flight plans, aircraft types, and mapping data;
- speed play of the created exercise;
- exercise speed up.

Within the ISS during the exercises preparation, the possibility for the advanced setting of the following shall be provided: exercise scope; plan-scenario of the exercise; weather conditions data and changing; messages forwarded to the controller; planned and flight data.

Information contained within the exercises database shall include definition of the aerodrome characteristics; setting and description of aircraft and ground vehicle types and characteristics; description of the static and dynamic graphical objects for the generation of the visual images; description of the geographic points; description of the holding routes; description of the air routes; description of the weather conditions including the visual objects visibility data; description of the navigation fields, radars, and radio direction finder characteristics; description of the flight routes; description of the standard SID/STAR trajectories.

Within the ISS, the following shall be implemented: dynamic change of the exercise parameters and monitoring of the exercise process (start, stop, pause, continue); simulation of the radar, ADS receivers, GNSS, direction finder, and communication malfunctions/failures following the instructor's commands; the possibility for the instructor to command for simulation of emergencies and in-flight emergencies. Within the ISS, measurement and record of certain parameters shall be provided during the exercise with their display on the instructor's screen. After the completion of the exercise, an integral assessment of the work of each controller–trainee shall be generated automatically.

Architecture and Administration Requirements The ISS architecture shall be based on the open modular system with the distributed structure and data processing based on the LAN, which means that it shall be composed of a number of workstations, work pairs, which are built around the LAN, and it shall provide for the possibility to increase the number and further modify not only the tasks to be solved by the STD, but also the data display aids. During the ISS development, the possibility to increase the level of automatization, changing of the efficiency and tool capacity of the ISS, the possibility to connect new data sources, modernization or substitution of the technical aids and software components with mode advanced equivalents. Within the STD, the possibility to provide the exercise configuration based on the

STD modules shall be provided: single module, which operates all workstations of the LAN; several modules, which are each operating by its own scenario.

The STD module is a network segment, which operates independently by a separate training scenario.

Within the ISS, the possibility to configure the STD modules shall be provided. The configuring functions shall be provided through software from the workstation of the administrator or training supervisor.

The modules shall be configured in the following manner: the modules shall consist of the AWSs of controllers and AWSs of operator pilots/instructors; the minimum composition of the module: one AWS of operator pilot/instructor and one AWS of the trainee (controller); within a module, the number of the workstations shall be increased one by one, at that the combination of the numbers of the AWSs of controllers and instructors shall be optional.

The operation of separate modules shall be independent and have no influence on the operation of other modules. During the modules configuring access to the database server shall be arranged in order to create the exercise scenario or to choose from the list of previously created exercise scenarios. During the modules configuring the functions of each workstation, control area sectors shall be defined, as well as the respective voice communication. All data created within the module, the process of the performed exercise and its results shall be recorded and archived in accordance with the respective requirements for one of the AWSs of the training supervisor or instructor within the module, as well as subsequently stored in the common databases of the STD.

The training supervisor and system engineer/administrator shall have the possibility to connect to any STD module in order to monitor the exercise performance without any preliminary request and confirmation, but having no influence on the module operation.

Section Review

1. Main designation requirements set for the STD of the ATC AS controllers.
2. Main tactical requirements set for the STD of the ATC AS controllers.
3. Main technical requirements set for the STD of the ATC AS controllers.
4. Training requirements set for the STD of the ATC AS controllers.
5. Architecture and administration requirements set for the STD of the ATC AS controllers.

References

1. Filin AD et al (2001) Development of a technology for constructing a simulator-modeling package for practicing joint operations of flight crews and flight management services in a complex environment. Collection of abstracts. In: International conference "training technologies and training: research, development and market needs". Zhukovsky, TsAGI
2. Filin AD, Suleymanov RN et al (2011) Integrated simulator for training flight operations directors. Explanatory note for the technical project. VNIIRA, SPb., p 250. Inv. 27262

Chapter 5
Architecture and Composition of ATC Automation System Simulators for Controllers

5.1 Structural Scheme and Composition of Integrated System Simulator (ISS)

The structure of an ISS must strictly correspond with the organizational structure of the simulated ATS region, and in accordance with the Federal Aviation Rules (FARs), the number of ATC sectors and ATC controller workstations in an ATS unit structure is defined based on the provision of the acceptable level of flight safety in an ATS airspace [1, 2]. The architecture of the advance ISS is composed of open modular information systems that are based on the local area network and with distributed data processing structure.

The structure of modern ISS is based on the following main technical and organizational principles [3]: modularity; openness of the architecture; structural versatility; expandability of the structure subject to the user's demands; invariance of the configuration; possibility of changing the functional purpose of the modules; high level of adequacy of simulation of the air situation and radio and technical aids; high level of automatization of the functions of operating/support staff.

Functionally, the ISS consists of two main subsystems: training subsystem and simulation and assessment subsystem. The ISS architecture is shown in Fig. 5.1. The training subsystem is comprised of four groups of universal automated workstations of the trainee–controllers (AWS-C):

– AWS-C for radar monitoring 1…N;
– AWS-C for procedural monitoring 1…M;
– AWS-C for planning 1…L;
– AWS-C for visual monitoring 1…K (runway supervisor, taxiing supervisor).

The simulation and assessment subsystem comprises:

– AWS-OP 1…K;
– AWS-TS;

© Springer Nature Singapore Pte Ltd. 2020
Bestugin A.R. et al., *Air Traffic Control Automated Systems*,
Springer Aerospace Technology, https://doi.org/10.1007/978-981-13-9386-0_5

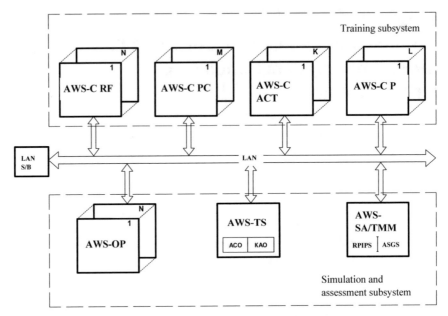

Fig. 5.1 Structure and composition of the complex system simulator

– AWS-SA/TMM (it is possible to arrange combined or separate AWS for the system administrator and technical management and monitoring—depending on the complexity of the ISS composition).

All AWSs of the ISS are connected through a local area network switchboard (LAN S/B) into a single local area network (LAN). The qualitative composition of ARW-Cs is defined by a set of design specifications of a specific simulator's modification. The quantitative composition of ARWs (N, M, L, K) is defined by the structure of the simulated airspace and by the ISS functional tasks.

With the help of AWS-TS, the automated training system (AutoTS) is launched, and the display of the results of the automated assessment unit (AAU) operation is provided. The ISS presented as a general structure in Fig. 5.1 can function in the following main modes: preparatory; operating; concluding.

In the preparatory mode, the operating/support staff conducts preparation checks on the exercises that will be played in real time in the operating mode. The required structure of the airspace area connected to the simulated aerodrome is selected from the structure library (if needed, a new model of a preset structure of the airspace area is created). The prepared exercise is tested in a fast-time scale. The exercises are prepared in accordance with the controllers training plan. In the operating mode, the exercises chosen from the exercise library are played in real time in accordance with the training sessions plan. During the training, the trained controllers' automated assessment unit is operating. In the concluding mode, the training debriefing takes place. The grading table with the grades received by each trainee is displayed; also, individual grading tables with incorrect or inadequate actions of a certain trainee can be displayed. At the same time, if required, some parts of the exercise can be played starting from any chosen point of time; the playback can be stopped or restarted. During the playback, correct commands can be given instead of the incorrect, and the consequences of their input can be studied.

5.2   Configuring of the STD Modules

In the ISS structure presented in Fig. 5.1, it is possible to configure the STD modules that consist of AWS-C and AWS-OP. The total number of the workstations and the number of AWS-C/AWS-OP combinations in each module shall be optional. The STD module is a segment of the network that works independently by a certain training scenario, and at the same time provides the playback of an independent exercise. The modules are configured to match the structure of the simulated airspace in a manner described below. AWS-C and AWS-OP are included into modules; the minimum module composition is one AWS-OP and one AWS-C. Within a module, the number of the workstations can be increased one by one, in that the combination of the AWS-C/AWS-OP numbers is optional. Thus, the configurations are arranged in such a way to allow one AWS-C to function with several AWS-OPs when difficult exercised are played, and vice versa, to allow several AWS-Cs to function with one AWS-OP when simple exercises are played. In that simultaneous training, several groups of trainees are arranged in a separate group trainings mode.

At that, each separate module works independently following its training scenario and does not influence the operation of other modules. If this requirement is fulfilled, it significantly broadens the training capacity of the ISS. When such configuring function is available, it allows training the controllers of area control centers and aerodrome control towers (area, approach, circuit, landing, runway, taxiing) with the possibility to arrange both individual controllers' training by sectors and integrated training—simultaneously for a connected combination of the controllers of the said ATC sectors. Initiation and configuring of the STD modules are provided through AWS-TS with the help of special dialog panels: "Launch" and "Configure." During the module initiation, the system addresses the database of previously prepared exercise scenarios. The training supervisor chooses the required scenario and assigns control sectors and workstation types for each workstation that is included into this module.

After the end of the initiation and configuring process, the module switches to the operating mode. During the whole time of operation of the STD module, the supervision information on the condition of the workstations is displayed at the training supervisor's display. The training supervisor can choose to display the current condition of the ASD (ground situation display at the aerodrome in the plan), the current condition of the phraseology windows, and the coordination with the adjacent centers of any instructor that participates in the training process. The training supervisor can also display the following information windows: meteorological condition window, lists of active and buffer aircraft, and current aircraft parameters. When several STD modules are launched, the information is displayed in several dialog windows.

The training supervisor can see the required information by his/her choice in order to monitor the training process within each launched exercise.

The configuring and control interface provides for:

- choosing the exercise from the exercise library for training, where each exercise is required for specific training purposes and describes the specific airspace area with the indication of the planned flight table;
- choosing the control area, in that the list of the control areas (areas of responsibility) is displayed, where this list was prepared during the description of the airspace area; this possibility is used when the prepared exercise is fine-tuned;
- choosing the options that allow adjusting the STD operation parameters:
- AWS-OP display color settings: choosing the colors for the display of the menu, reports, notifications, current parameters, aircraft under control and located in the adjacent area, etc.;
- choosing the control modes: allows choosing manual (to control each aircraft) or automated (used to fine-tune a newly created exercise, where each aircraft follows the flight path described in the plan) mode;
- choosing the additional control mode to allow for the inputs from the absent adjacent control area (where the aircraft appears some time before reaching the control area border), or lack of such additional control mode;
- launching the simulator;
- requesting the report: allows receiving information on the work of the software (simulator launching, ways to input the control commands);
- exit: stops the operation of the simulator.

Section Review

1. Structure and composition of the integrated system simulator for the controllers of the ATC AS.
2. Modes of operation of the integrated system simulator for the controllers of the ATC AS.
3. Configuring the workstations of the integrated system simulator for the controllers of the ATC AS.
4. Composition and functions of the STD module of the controllers of the ATC AS.

References

1. Federal Aviation Regulations "Flights in the airspace of the Russian Federation" (2002) Transport, Moscow, p 96
2. Federal Aviation Regulations "Civil aviation flights" (2005) Military Publishing House, Moscow, p 224
3. Filin AD, Suleimanov RN et al (2005) Development of proposals on the formation of the technical appearance of the automated training system and integrated simulators for training flight operations directors and officers of the combat control of the ship and ground centers of aviation control. Report on R&D. VNIIRA, SPb., p 150

Chapter 6
ISS Automated Workstations

6.1 Training Supervisor's Automated Workstation

The (ISS) integrated system simulator includes an automated workstation of the training supervisor (AWS-TS) that enables accomplishment of functional tasks regarding management, monitoring, and subsequent analysis of the training process [1].

The structure of the training supervisor's automated workstation (AWS-TS) consists of two sections: the AWS-OP section and the AWS-C RF section.

The AWS-TS structure is shown in Fig. 6.1. The AWS-TS AWS-OP section interface provides access to the information necessary for setting up, configuring, and operating the simulator. It provides the capability to enter and edit the database of the simulated airspace structure and exercise scenarios. When an exercise scenario is prepared, one can view the progress of the exercise in the "fast-forward" mode (or in real time) at this workstation automatically. Viewing the progress of the exercise makes it possible to more accurately simulate the temporal parameters of the air situation scenario to solve specific training tasks. The AWS-TS AWS-OP section interface enables configuration of a single training module or several individual training modules, selection of the necessary exercise scenario, and launch of the training process. During the training, one can enter commands to complicate or simplify the air traffic situation, enter adverse weather conditions, enter additional simulated airplanes, set special in-flight cases, and create conflict situations.

To monitor the progress of each exercise launched, a window is displayed on the screen of the AWS-OP section that contains current information on the status of the workstations included in a specific training module. On the screen of their workstation, the training supervisor can display the necessary information windows and the ASD window provided for in the AWS-OP interface. From the AWS-OP section, the training supervisor can interrupt (stop) the training process, continue the training, and complete the training. The automated training system (AutoTS) can be launched from the AWS-TS. At their workstation, the training supervisor can call the

© Springer Nature Singapore Pte Ltd. 2020
Bestugin A.R. et al., *Air Traffic Control Automated Systems*,
Springer Aerospace Technology, https://doi.org/10.1007/978-981-13-9386-0_6

Fig. 6.1 Structure of the automated place of the head of the training

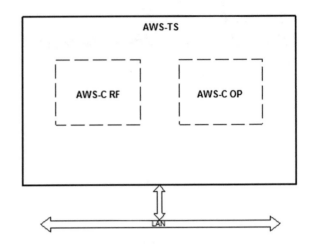

results of the automated evaluation complex performance, request error recording tables of the trainees to analyze the training process. The display and control means of the AWS-C RF section provide the training supervisor with an appropriate model of the air situation similar to the AWS-C RF of the training subsystem.

The training supervisor can address the database of the documentation and replay complex, select the desired recording, and start the replay process of the previously completed exercise. The exercise will be played on those AWSs of instructors and controllers that were included in the training module during the recording process. The AWS-TS provides the display and implements the human–machine interface of a given workstation of the air traffic controller that is simulated in the ISS.

6.2   System Operator/Pilot's Automated Workstation (AWS-OP)

The AWS-OP is built on the basis of a workstation with a liquid-crystal display, standard input controls, and a touch screen for displaying a simulated control panel for ground-to-air voice communication channels and service loud-speaking communication. The AWS-OP is designed to organize the management of the air traffic situation and control over the actions of air traffic controllers. The main functional task of the AWS-OP is to simulate actions of the crews of aircraft that are under control in the corresponding ATC area. The AWS-OP provides issuance of necessary messages to the AWS-C in the training process and exercise of the controller's commands to control the aircraft in the control zone over a given simulated radio channel. During the exercise, the following is provided:

– simulation of a controlled air traffic situation and transmission of relevant information to all workstations of air traffic controllers being trained;

- controllers trained control the movement of the simulated aircraft by issuing commands using standard radio phraseology;
- all the information about the actions of air traffic controllers trained is recorded;
- development of individual indicators of automated evaluation of the actions of the controllers trained based on the main criteria and parameters for ensuring flight safety in ATC systems;
- the instructor can intervene in the course of the exercise (enter (PCSs) potential conflict situations, enter additional aircraft, change meteorological conditions, stop and start exercises, etc.);
- simulation of meteorological data including hazardous weather conditions and runway weather.

The following functions are provided to control aircraft from the AWS-OP:

- transfer of control between AWS-OPs;
- taxiing and stopping the aircraft moving on the aerodrome maneuvering area according to commands and routes appointed by the air traffic controller, as well as during exercises involving evacuation of aircraft from the airfield;
- control of arriving and departing aircraft according to commands and routes assigned by the air traffic controller;
- setting an altitude (flight level);
- setting the vertical rate of climb/descent;
- setting an altitude (flight level) to a specified point;
- setting parallel displacement;
- setting a heading;
- setting a turn at a certain angle;
- setting a bank angle when turning;
- setting the direction of the turn;
- setting a new code of the transponder;
- simulation of the transponder modes;
- flying to a specified point;
- flying to the holding area and flight in the holding area;
- changing the flight route;
- dispatching aircraft following SID/STAR procedures.

The AWS-OP structure is shown in Fig. 6.2 and includes the means of simulated radio communications (MSRC), means of simulated service loud-speaking communication (MSLC), and AWS-OP (AOP) display with information processing means.

The AWS-OP screen displays the information control field (Fig. 6.3) implemented on the basis of the standard window interface.

The main elements of the AOP information field are the air situation display (ASD) window and a set of windows and panels to set conditions and enter control commands [1]. There are windows provided for selecting the training mode, adjusting the colors, adjusting the fonts, adjusting the display parameters, and selecting the zone patterns.

Fig. 6.2 Structure of system
operator pilot's automated
workstation

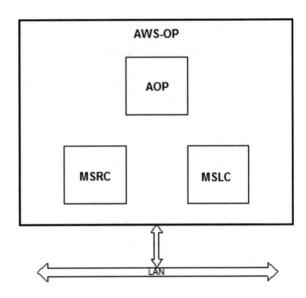

The training mode can be manual and automatic. It is possible to set or reject the planned flights. The color settings menu allows customization of the background color for the operator pilot, instructor, AWS-OP ASD screens, and their elements (reports, messages, etc.) or selection of one of the three previously defined color solutions taking into account the ICAO ergonomic requirements and recommendations. The display settings menu allows customization of the display parameters defining the image displayed on the ASD.

The zone pattern selection menu makes it possible to add a picture from the list of available patterns or delete a picture from the ASD window.

At the upper part of the AOP, below the heading string, a menu bar is displayed with a list of operator pilot and training supervisor's functions. To switch the menu from the main mode of the operator pilot's workstation to the training instructor/supervisor mode and back, one can use an icon on the right part of the screen, "I". Icons "K" and "S" are designed to display prompt messages on the key assignment and selection of setting items, respectively.

The left part of the screen constantly displays a window with lists of flight (aircraft) numbers.

When an aircraft is activated, the screen displays windows of the current aircraft parameters, the flight level panel, and the digital dialing unit to the right of the list of flight numbers. The lower part of the screen constantly displays lines for the reports from the aircraft under control and the aircraft from the buffer zone.

The AWS-OP is multi-purpose and can be used for training the controllers of all the sectors:

– area;
– approach;
– circle;

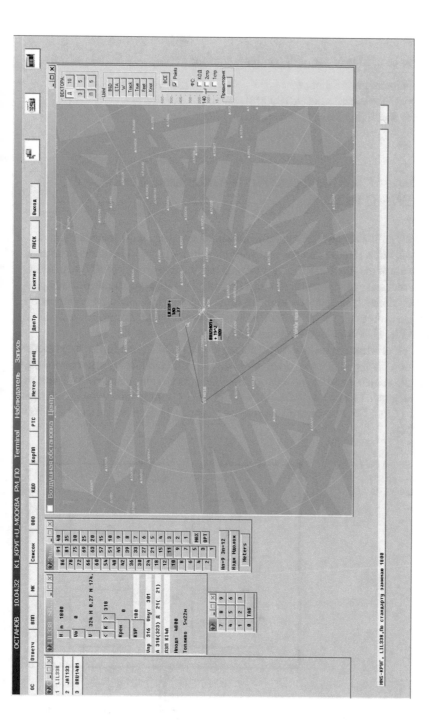

Fig. 6.3 Example of displaying information on operator pilot/instructor's screen (in Russian)

– runway (landing)/start;
– taxiing,

It can also be used in certain ISS configurations to:

– provide instructor's functions;
– "team-up" for other (adjacent) control centers;
– perform the flight planner's functions.

6.3 Radar Facilities Controller's Automated Workstation (AWS-C RF)

The radar facilities controller's automated workstations are implemented in the ISS on the basis of a remote control unit with built-in hardware. The main functional components of the AWS-C RF are a synthetic air situation display (ASD) with information processing equipment; reference data display (RDD) with information processing equipment; means of simulated ground-to-air voice radio communication (MSRC); means of simulated service loud-speaking communication (MSLC).

The AWS-C RF structure is shown in Fig. 6.4.

Sources of information that is displayed on the ISS RF controller's synthetic air situation display and used as the basis of air traffic monitoring and control are primary surveillance radar; secondary surveillance radar; automatic radio direction finder; integrated radar data (IRD); ADS-B and ADS-C information; controller–pilot data link (CPDLC); data transmission links; planning information; meteorological information.

Fig. 6.4 Structure of radar facility controller's automated workstation

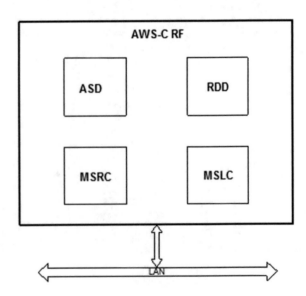

The air situation display shows the result of processing from the information sources, in accordance with the capabilities of the simulated ATC system. The trained controller's interface for managing the information displayed on the air situation display fully corresponds to the functionality of the simulated ATC system. As a rule, the ISS uses an application software package of a given ATC system to ensure the proper functioning of the AWS-C RF. In the air situation generator server (ASGS) (Fig. 6.1), information sources are simulated that are used in the ATC system simulated in the ISS. The radar, navigation and planning information processing server (RPIPS) (Fig. 6.1) operating on the basis of the application software package of the simulated ATC system provides information for the synthetic air situation displays in the ISS AWS-C RF. The reference data display shows current information, both constant and variable, on the aeronautical situation, meteorological situation, air traffic flow management planning, the state of technical facilities, and information sources to support flight control. The list of information displayed on the reference data display and the control interface corresponds to the simulated ATC system. Storage and issuance of reference data in real time are provided by means of the ASGS databases generated during the ISS preparatory mode of operation. Means of simulated ground-to-air voice radio communication include the control and ground-to-air radio communication channel switching panel, which is implemented on the basis of a touch screen display, and a set of terminal devices (remote control PTT switch, foot PTT switch, microphone, and handset). Means of simulated service loud-speaking communication include a panel for selecting subscribers for loud-speaking communication on the basis of a touch screen display and terminal devices for voice loud-speaking communication.

6.4   Procedural Air Traffic Controller's Automated Workstation (AWS-C Pro)

Procedural air traffic controller's automated workstations are included in the ISS structure for configurations implementing simulation of the Unified ATM System ACs and LSUs. ISS AWS-C Pros are implemented on the basis of a remote control unit with built-in hardware. The main functional components of the ISS AWS-C Pro hardware are

- synthetic air situation display (ASD) with information processing equipment;
- reference data display (RDD) with processing equipment;
- means of simulated service loud-speaking communication (MSLC).

The AWS-C Pro structure is shown in Fig. 6.5.

The main elements of the AWS-C Pro are the air situation display (ASD), the reference data display (RDD), and means of simulated service loud-speaking communication (MSLC).

Fig. 6.5 Structure of the
procedural air traffic
controller's automated
workstation

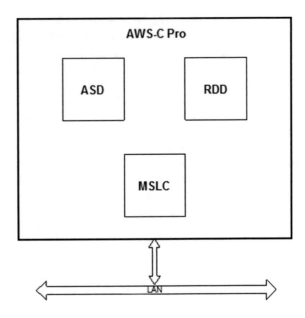

Sources of information shown on the synthetic air situation display for the ISS Pro controller provide information that are used by the procedural air traffic controller who performs their functions as an RF controller's assistant.

The air situation display outputs the result of processing from the information sources, in accordance with the capabilities of the simulated ATC system. The air situation display shows aircraft routes.

Ways to display information about the location of simulated airplanes, creation of the controller's schedule of the aircraft routes, displaying the vertical movement profile, displaying information about compulsory reporting points with time and altitude of their overflight, displaying auxiliary information (icons confirming the arrangement to leave the ATS area, hazardous weather, prohibitions, restrictions, etc.) shall correspond to the information support of the automated ATC system modeled in the ISS.

The trained controller's interface for managing the information displayed on the air situation display fully corresponds to the functionality of the simulated ATC system. As a rule, the ISS uses an application software package of a real ATC system to ensure the proper functioning of the AWS-C Pro. In the ASGS (Fig. 6.1), information sources are simulated that are used in the ATC system simulated in the ISS. The RPIPS (Fig. 6.1) operating on the basis of the application software package of the simulated ATC system provides information for the synthetic air situation displays in the AWS-C Pro.

The reference data display shows current information, both constant and variable, on the conditions of flight operations in a given ATC area. The list of information displayed on the AWS-C Pro reference data display and the control interface corresponds to the simulated ATC system and is identic to the AWS-C RF information support. Storage and issuance of reference data in real time are provided by means of the ASGS databases generated in the preparatory mode of ISS operation.

6.5   Planning Controller's Automated Workstation (AWS-C P)

Planning controller's automated workstations (AWS-C P) are included in the ISS structure for configurations implementing simulation of the full process cycle of air traffic management and control. AWS-C Ps are realized on the basis of a universal remote control unit with built-in hardware. The main functional components of the AWS-C P hardware are planning information display (PID) with information processing equipment; reference data display (RDD) with information processing equipment; means of simulated service loud-speaking communication.

The AWS-C P structure is shown in Fig. 6.6.

The ISS functionality provides for the simulation of all information messages, the formation and maintenance of special service information to ensure the implementation of the AWS-C P operational functions for the simulated ATC system exercises involving their procedures, including the following:

– information on flight plans and ATC messages;
– meteorological information;
– aeronautical information and information on restrictions on the airspace use;
– reference and auxiliary information;
– formation and maintenance of the daily flight plan;

Fig. 6.6 Structure of planning controller's automated workstation

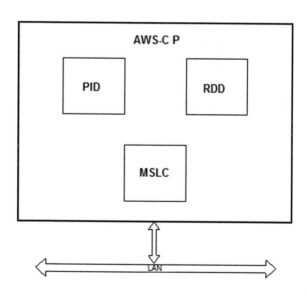

- formation and simulation of outgoing messages over the Civil Aviation Terrestrial Data Transmission and Telegraph Communication (TDTTC) network;
- information flow of incoming TDTTC network messages;
- formation of output forms;
- printing messages and output forms;
- control of the availability of licenses for flight operations;
- calculation of the projected load of ATC sectors and areas;
- calculation and visualization of the planned flight routes;
- drawing up, analyzing, and maintaining the daily flight plan;
- generation and issuance of standard airspace use messages in accordance with the message table;
- monitoring and analysis of incoming airspace use messages;
- monitoring and analysis of the meteorological situation at the associated and alternate aerodromes, as well as on the routes and in the areas of flight operations;
- monitoring and analysis of aeronautical information on the status of aerodromes, RTF, and air routes;
- interaction with adjacent control centers and airport services.

The reference data display shows the current and future planning information. The planning data display provides the formation, input, and monitoring of current messages regarding the planning controller's operations technology. The planning controller's interface of the operations procedures corresponds to the procedures of the simulated ATC system. Preparation, creation (in accordance with the exercise plan) of the planning information generated during the ISS preparatory mode of operation, its storage, and real-time issuance during the exercise are provided by means of the ASGS databases.

6.6 Aerodrome Control Tower (ACT) Controllers' Automated Workstations

6.6.1 Final Controller's Automated Workstation (AWS-C F)

The final controller's automated workstation is included in the ISS structure for configurations implementing simulation of the ACT responsibility areas. The main functional components of the AWS-C F equipment are as follows:

- synthetic air situation display (ASD) with information processing equipment;
- landing radar simulator display (LRSD) with information processing equipment;
- airfield control radar simulator display (ACRSD) with information processing equipment;
- simulated lighting equipment control panel (SLECP) screen with information processing equipment;
- means of simulated "ground-to-air" voice radio communication (MSRC);

Fig. 6.7 Structure of the
final controller's automated
workstation

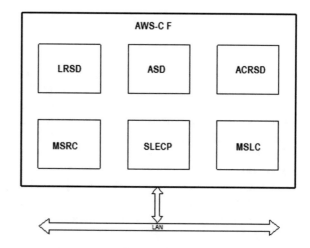

– means of simulated loud-speaking communication (MSLC).

The AWS-C F structure is shown in Fig. 6.7.

Sources of information that is displayed on the ISS controller's synthetic air situation display and used as the basis of air traffic monitoring and control are models of radar equipment listed in this section.

Sources of information displayed on the LRSD and ACRSD that are the basis of the final controller work procedures are landing radar and airfield control radar models. The air situation displays, as well as the landing radar and the airfield control radar displays, output the results of processing from information source models, in accordance with the capabilities of the simulated ATC system landing area.

The trained final controller's interface to manage the information displayed on the ASD, LRSD, and ACRSD fully corresponds with the functional simulated radar systems in accordance with the capabilities of the simulated ATC system.

In the air situation generator server (ASGS) (Fig. 5.1), information sources are simulated that are used in the ATC system simulated in the ISS.

The radar, navigation and planning information processing server (RPIPS) (Fig. 5.1) operating on the basis of the application software package of the simulated ATC system provides information for the ASD, LRSD, and ACRSD. The simulated lighting equipment control panel screen is implemented on the basis of a touch LCD display. The mnemonic scheme and the control interface correspond to the lighting equipment of the simulated aerodrome.

6.6.2   Runway and Taxiing Controller's Automated Workstations

Aerodrome control tower (ACT) controllers' automated workstations, runway supervisory unit (RSU), and taxiing supervisory unit (TSU) are included in the ISS structure for configurations implementing simulation of ACT responsibility areas. The main functional components of the AWS-C F hardware are ASD; ACRSD; PDD; simulator of the visual airfield conditions (SVAC); MSRC; MSLC. The AWS-C RSU (STU) structure is shown in Fig. 6.8. The hardware included in the AWS-C RSU (STU) is incorporated into a local area network. The ISS SVAC provides modeling and simulation of AWS-C RSU (STU) visual environment in the area of the simulated aerodrome. Air and ground situations are modeled in three dimensions (3D modeling) and displayed on specified display media (plasma panels, liquid-crystal displays, projection systems) in the form similar to the view from the controller's workstation located on the tower of a particular aerodrome, and also in specified points in the territory of the airfield. The basis for the creation and operation of the visual aerodrome conditions model are maps, charts, photographic images, and other similar sources (space and aerial photographic images) that make it possible to model an aerodrome, its surrounding area and ground facilities with sufficient detail to create the required panorama.

The coverage of the territory to be modeled for visualization shall correspond to the minimum angle of view from the modeled ACT controller's observation point, and to the distance up to 15 km. A digital model of the terrain of a modeled aerodrome is created based on the use of a stereo pair of satellite images or using large-scale geographical maps with a horizontal increment of 20 m maximum. Fixed objects at

Fig. 6.8 AWS-C RSU (STU) structure

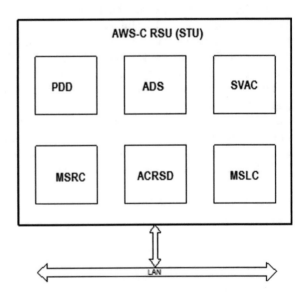

the airport are modeled on the basis of data from large-scale schemes, plans, and aerial photos of the relevant territories on the basis of relevant documents. Moving objects are implemented in the form of models with a given degree of detail. It is possible to use both libraries of models created by the simulator (visualization system) manufacturer, and libraries of other developers if they meet the requirements of this document. The main static objects of modeling are

— the surrounding area with terrain detailing, reference natural and artificial landmarks visible from the observation point (mountains, hills, masts, pipes, panorama of settlements, power transmission line poles, etc.);
— the territory of the airport with one or more runways (1–3);
— aprons, parking (aircraft stands), taxiways, landing areas;
— buildings and constructions;
— equipment (facilities) of radiotechnical flight support (RTS), meteorological equipment;
— other objects typical of the simulated aerodrome.

The main dynamic objects of modeling are

— aircraft (in accordance with the requirements of the simulated aerodrome, but not less than 25 types);
— surface transport, special equipment, and apron mechanization facilities operated at civil airports (in accordance with the capabilities of the simulated aerodrome, but not less than ten types);
— people (single persons, group);
— animals (single animals, herd);
— birds (flocks);
— other objects that are typical of the simulated aerodrome.

The SVAC software package enables real-time synthesis and reproduction of the visual situation in the entire field of view (with a refresh rate of at least 20 frames per second). A special library of programs as part of the SVAC provides management of the following parameters of the visual environment. Light control: the visual environment is displayed taking into account changes in the time of day (the current time set), as well as the seasons (winter, summer); control of the position of stars and celestial bodies in accordance with the set current time.

Management of meteorological phenomena: setting the place and intensity of clouds with the ability to control their movement depending on the direction and magnitude of the specified wind (up to three layers); a general change in the visibility range from 0 to 15 km; fog (general, local, surface); precipitation (rain, snow) with several degrees of intensity.

Management of dynamic objects: movement of airplanes and helicopters in the air in the takeoff and landing areas; movement of airplanes, helicopters, and surface transport in the territory of the aerodrome including aircraft following-up and towing; simulation of ailerons, flaps, airplane propellers, and helicopter main rotor systems.

Management of special effects of the visual environment: simulation of the aerodrome lighting equipment and aircraft lights; explosions; fire of varying intensity;

smoke of varying intensity with shifts depending on the direction and magnitude of the specified wind.

When simulating the route of moving dynamic objects at the aerodrome, the rules and regulations established by the governing documents at the aerodrome (normative speed, request for departure to the airfield, compulsory communication sessions, reports on clearing the airfield, etc.) shall be observed. For a detailed view of individual sections of the aerodrome and airspace, the SVAC provides views with at least tenfold magnification (the binoculars function). Thus, at ISS ACT AWS-Cs (TSU, RSU, F), the following main functional tasks shall be solved:

- displaying the results of the simulation of multi-radar processing of radar data and ADF information;
- modeling and displaying radar and visual information on the movement of aircraft and airfield equipment on the maneuvering area, the final approach track of a particular aerodrome and its airside;
- processing of planning information at the stages of preliminary and current planning and management;
- modeling of the planning information management process;
- predicting the air situation on the basis of planning information;
- analysis of the predicted and current air situation on the basis of planning and radar information regarding the absence of conflicts and detection of the following;
- achievement of the limits of the separation standards;
- potential near-midair collisions;
- PCSs with airspace use restrictions and hazardous meteorological phenomena;
- aircraft non-compliance with the meteorological landing conditions (the degree of automation in solutions for these problems is determined by the capabilities of the simulated ATC system);
- automatic generation of warnings regarding extreme deviations from the heading and the glide path on the final approach track;
- combined display of information on the air situation and planning data;
- displaying information on the runway state;
- displaying and control of the artificial runway simulation: "FREE-OCCUPIED";
- displaying information on the RTS equipment;
- processing and combined display of meteorological information;
- control of runway lights and taxiway lights from the final controller's workstation.

Section review

1. Configuration and main functions of the training supervisor's automated workstation.
2. Configuration and main functions of the operator pilot's automated workstation.
3. Configuration and main functions of the radar facilities controller's automated workstation.

4. Configuration and main functions of the procedural air traffic controller's automated workstation.
5. Configuration and main functions of the planning controller's automated workstation.
6. Configuration and main functions of the final controller's automated workstation.
7. Configuration and main functions of the runway and taxiing controller's automated workstation.

Reference

1. Filin AD, Suleymanov RN et al (2003) Upgrade of the ATC system training module "Spectr-M". Explanatory note for the technical project. SPb.: VNIIRA, p 131

Chapter 7
Organization of Simulated "Ground–Air" Radio Communication and Service Loud-Speaking Communication

Means of simulated voice communication (MSVC) are intended to simulate two-way radio communication between the AWS-C and AWSs of operating personnel through one of the selected radio channels, as well as for loud-speaking communication (LSC) between any ISS AWSs. To implement the ISS functionality, the MSRC and MSSC must provide the following:

– communication management using a touch screen displaying the control panel;
– capability to quickly select the operational mode;
– establishment of communication in any mode;
– voice communication in the microtelephone and loud-speaking modes;
– capability to select several channels for simultaneous listening with separate volume control;
– visual indication of operational modes and current connections on the touch screen;
– initiation of simulated bearing when conducting radio communication;
– two-way simplex radio communication between the ARM-C and the ARM-OP;
– capability to select any of the simulated radio channels to control;
– LSC between any workplaces with visual and audible indication of a subscriber's call.

For the organization of simulated voice communication in the ISS, the most preferable method is that based on the representation of MSVC as subscribers of a local computer network. This method is the most preferred, since it is based on the use of modern principles of information technology. The structure of the voice communication organization in the ISS presented in Fig. 7.1 is homogeneous and with open architecture.

Means of simulated voice communication (MSVC) are implemented on the use of sound cards in MSRC (MSSC) PCs at FSS AWSs. Quantitative values N and M are determined by the composition and configuration of the modeled ATC system structure. Analog information of voice traffic coming from AWS microphones is converted by the ADC into digital code transmissions to the corresponding LAN subscribers. In MSRC (MSSC) PCs of the receiving subscribers, the reverse DAC of the received digital transmissions takes place producing analog voice traffic signals

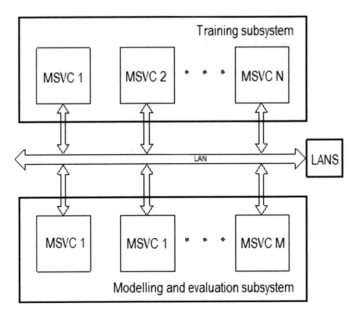

Fig. 7.1 Structure of simulated voice communication in the ISS

with subsequent output to the loudspeakers. The ARM MSVC equipment contains the following: workstation; touch screen; microphone; loudspeakers; handset/headset (H/H); connection panel. At workstations, network and sound cards are used.

The voice data of the selected source are divided into fragments of 20 ms in length, which are placed in voice information packets and sent to the LAN. Multimedia data are quite voluminous by their nature; therefore, to reduce the amount of documented voice information, two main mechanisms are used: audio data compression by a voice codec and exclusion (elimination) of pauses (speech-offs) using a voice activity detector.

Operator's commands (e.g., to establish a connection with a given subscriber) are converted into control IP packets and transmitted over the same network as voice ones. The structure of the MSVC software is presented in Fig. 7.2.

The MSVC workstations software contains the following units: control panel display module (PDM); control command processing module (CCPM); channel memory module (CMM); DAC-ADC conversion module (ConvM); compression module (CompM); database documentation and replay module (DDM); TCP/IP exchange organization module.

For transmission, a switched Ethernet network (1000 Base-TX) can be used; the network configuration is determined by the ISS configuration.

For the normal functioning of the ISS, the following voice communication modes are implemented. Duplex communication: two-way communication, when both subscribers can simultaneously transmit and receive signals. A communication session is established and terminated once at the initiative of any subscriber (service loud-

Fig. 7.2 The structure of the
MSVC software

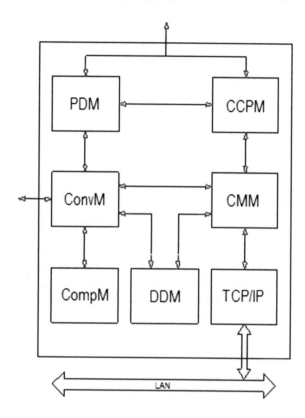

speaking communication). Simplex communication: one-way communication, when one subscriber transmits signals and the other one accepts them. A communication session is established and terminated at each transmission at the initiative of the transmitting subscriber (imitation of ground-to-air radio communication). Half-duplex communication: two-way communication, when subscribers can transmit signals in turn. A communication session is established and terminated once at the initiative of any subscriber. Example: communication via selective subscriber call (SSC) channels. Call forwarding: switching of the established connection or incoming call by one of the subscribers to another subscriber. Permanent forwarding: a mode in which all calls coming to this workstation are automatically transferred to another predefined workstation. Circular: communication, when one participant can speak (initiator of the conference), and the others can hear only him/her. Conference: connection of several subscribers, when all conference participants have the capability to simultaneously talk and hear each other. Holdup: a mode, in which a subscriber is temporarily "disconnected" without breaking the connection with the possibility of a subsequent re-connection. Intrusion: entering into communication with a subscriber engaged in another connection. Call pickup: an answer to a call coming to another workstation. Direct online access: the connection is established immediately after pressing the address key of the subscriber. The call is made by voice. Direct access: the subscriber receives a call signal when pressing the address key. Communication is possible only after the subscriber answers. Direct online access and direct access types from a particular workstation are available only for subscribers for whom address keys are provided. Radio channel taking for control: switching on the radio channel in the "control" mode. Radio transmission is possible in this mode only. The listening channel is a radio channel that only receives signals. There are listening channels over which transmissions are not impossible in principle. Also, a standard radio channel released from control becomes a listening channel. An available radio channel is a radio channel that can be monitored and controlled from this workstation. The number of available radio channels may exceed the number of radio communication keys on the touch screen. The user can assign a free key for any of the available radio channels.

The view of the MSVC control panel displayed on the touch screen is shown in Fig. 7.3.

The panel is divided into three main zones: on the left, there is a zone of function keys designed to activate various operational modes; in the center, there is a loud-speaking communication zone consisting of one or several address key panels; buttons in the upper part of the zone are intended for switching between panels; on the right, there is a radio zone consisting of one or more panels of radio communication keys.

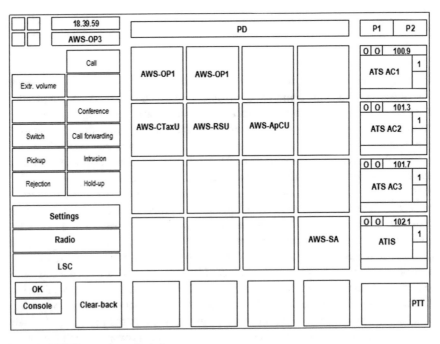

Fig. 7.3 MSVC control panel window

At the bottom of the screen, there are (from left to right): the clear-back button, the call panel, and the radio press-to-talk key—PTT. In the upper part of the function key zone, there is a system clock, a field displaying the name of the workstation (AWS-OP3 in the figure), and mode indicators. The absence of the figure means the ATIS module works offline. In the middle part of the function key zone of the panel, there are keys designed to activate various operational modes for the ATIS. The loud-speaking communication zone is represented as a panel of direct access to subscribers, which contains address keys, each of which displays the subscriber's name (The loud-speaking communication zone may have several direct access panels that are switched by means of buttons located in its upper part.). The current state of the subscriber is indicated by a change in the appearance (color) of the address key. Communication with the direct access subscriber is established by pressing the address key. The call panel is located at the bottom of the screen. The call bar buttons show incoming calls. The number of call panel keys is set by the administrator when configuring the workstation screen. In the absence of calls and connections, all the keys of the call panel are empty and painted gray. If there is a call, the key displays the subscriber's name and their internal number. The color of the key corresponds to the state of the call (communication), as well as for the address key in the loud-speaking communication panel.

If there is one call, the leftmost key of the call panel will be activated; if the second call comes, the next one is activated, etc. The radio communication zone contains one or several panels of radio communication keys and the press-to-talk key (PTT). Switching between panels is carried out using the keys ("P1", "P2") at the top of the panel.

The assignment of individual key fields and the appearance of radio communication keys assigned to the radio channel are shown in Fig. 7.4. The name for each radio channel is set by the system administrator during configuration. The number of the audio speaker indicates which of the speakers reproduces the signal from this group. If no radio channel is assigned to the key, the key is free. Free radio communication keys do not contain information.

The state of the radio channel is indicated by the key color or its individual elements (see Table 7.1).

The PTT key enables transmissions from all radio stations under control.

Setting the basic parameters of the system, such as setting up available loud-speaking communication subscribers and groups of radio stations, is performed by the system administrator and is not available at a particular workstation. Settings panels appear in the loud-speaking communication zone when buttons **Settings**, **Radio,** or **LSC** are pressed in the function key zone.

The general settings panel of a workstation is opened by pressing the "Settings" function key on the control panel (Fig. 7.3) in the loud-speaking communication zone of this control panel. The general settings panel is shown in Fig. 7.5.

The panel is designed to configure common control functions. The controls of the panel enable the following: turn on/off and restart the workstation software; adjust the replay volume. The "LSC" panel is opened by pressing the "LSC" button on the control panel (Fig. 7.3). The "LSC" panel is designed to control the loud-speaking

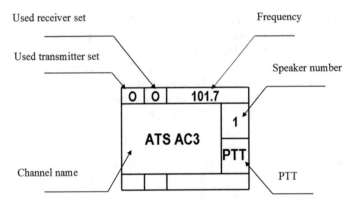

Fig. 7.4 Assigning radio communication key fields

Table 7.1 Key color or its individual elements

Key color	Channel operational mode	State of the radio channel or the monitored channel
Green	Control is ON, reception is ON	Free radio channel
Red	Control is ON, reception is ON	Occupied radio channel, the other subscriber transmits signals
Blue	Control is ON, reception is ON	Signal (carrier) present
Gray, text color is white	Control is OFF, reception is ON	Channel is off
Gray, text color is red	Control is OFF, reception is OFF	Channel is not available
Gray or yellow, frequency field is blue	Control is OFF, reception is not essential	Signal (carrier) present
Green or yellow, special functions string is blue	Control is ON, reception is not essential	The receiver of this channel is defined as the best in the group

communication, as well as to select the call signal. The controls of the panel provide the following features:

– selection of audio speakers for listening to loud-speaking calls;
– adjustment of the volume when listening to calls;
– selection and adjustment of the call signal volume.

The basic radio communication settings panel (Fig. 7.6) is activated by pressing the "**Radio**" function key on the control panel shown in Fig. 7.3.

The basic radio communication settings panel is shown in Fig. 7.6.

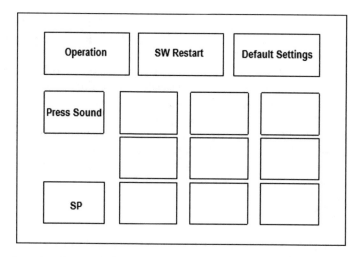

Fig. 7.5 General settings panel

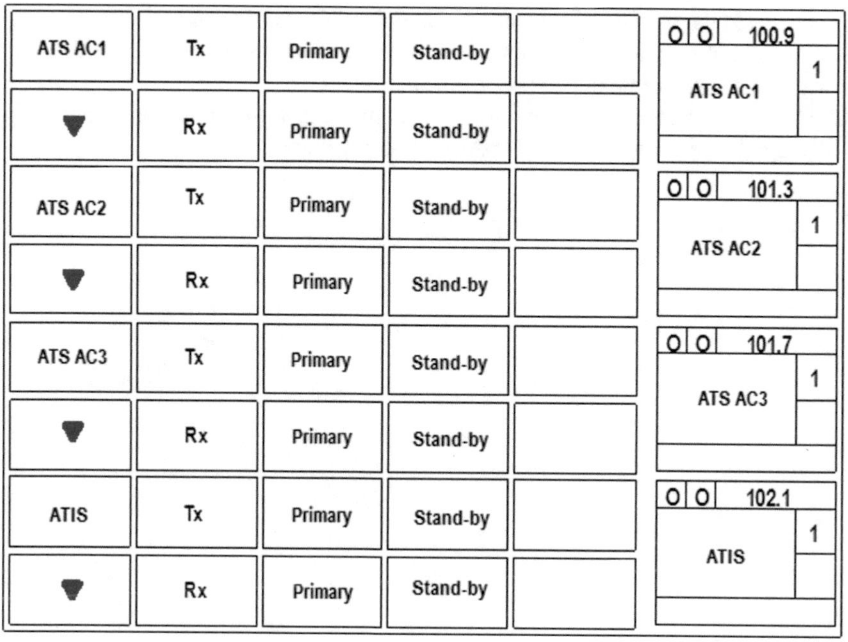

ATS AC1	Tx	Primary	Stand-by		O O 100.9 1 ATS AC1
▼	Rx	Primary	Stand-by		
ATS AC2	Tx	Primary	Stand-by		O O 101.3 1 ATS AC2
▼	Rx	Primary	Stand-by		
ATS AC3	Tx	Primary	Stand-by		O O 101.7 1 ATS AC3
▼	Rx	Primary	Stand-by		
ATIS	Tx	Primary	Stand-by		O O 102.1 1 ATIS
▼	Rx	Primary	Stand-by		

Fig. 7.6 Basic radio communication settings panel

The panel contains several groups of function keys. Each group is designed to configure the radio channel, the key of which is located to the right of the group in the radio communication zone.

The controls of the panels enable the following: turn on/off the radio channels for listening and/or control; select a primary or standby receiver/transmitter; activate the advanced radio communication settings panel for each channel.

The MSVC provide recording and replay of voice information for a given workstation. Recording and replay are performed synchronously by time stamps with the documentation on air and ground situation, which is performed in the server of the air situation generator. Voice data and air situation data recorded are stored at each workstation, for each exercise performed.

Section Review

1. Method and principles in ground–air–ground radio communication simulation means in the integrated system simulator for ATS AC controllers.
2. Method and principles in service loud-speaking communication simulation means in the integrated system simulator for ATS AC controllers.
3. Functional capabilities of voice communication simulation means in the integrated system simulator for ATS AC controllers.
4. Functions of the control panel for voice communication simulation means in the integrated system simulator for ATS AC controllers.

Chapter 8
Automated Training Aids for the Remote ATC Specialists Proficiency Maintaining System

8.1 Automated ATC Specialists Training and Assessment System

Nowadays, organization of theoretical training for air traffic controllers involves integrated automated training aids more and more widely; these aids operate in conjunction with remote proficiency maintaining systems (RPMS) built on the basis of modern PCs.

Consider one of the possible ways of assessing controllers' knowledge and skills using automated training aids, which is useful in the training of ATC specialists.

The functional diagram of the knowledge and skills assessment process in automated training aids is shown in Fig. 8.1. Here, the data files contain variants of tasks, arrays of initial data, parameters of the air situation, and corresponding standard performance models (standard values of monitored parameters). According to the assessment results, files are formed that are then used to analyze the training process. Criteria for assessing the level of practical skills are selected based on the type of actions performed by trainees. The selection of criteria determines the assessment algorithm, which is reduced to checking whether the monitored parameters belong to the area of acceptable (successful) solutions or actions.

Figure 8.2 shows the main tasks implemented by RPMS integrated automated training aids methodological support and software for ATC specialists. These include building a knowledge domain model (knowledge base); determining a list of advancement questions (tests, tasks); defining knowledge assessment criteria; selecting an assessment mode (number of test tasks presented); determining the level of requirements for the trainee in the assessment process; determining weights for questions that have appropriate evaluation criteria; implementing a knowledge assessment program on PCs (using various modes); developing an indicator to assess the level of the controller's knowledge; final assessment of the specialist's proficiency level (binary, N-points); viewing wrong answers of the trained controller; assessing the reliability of the assessment results; making decisions on the further organization of assessment (change of strategy or assessment tactics, termination of the assessment); drafting a

© Springer Nature Singapore Pte Ltd. 2020   185
Bestugin A.R. et al., *Air Traffic Control Automated Systems*,
Springer Aerospace Technology, https://doi.org/10.1007/978-981-13-9386-0_8

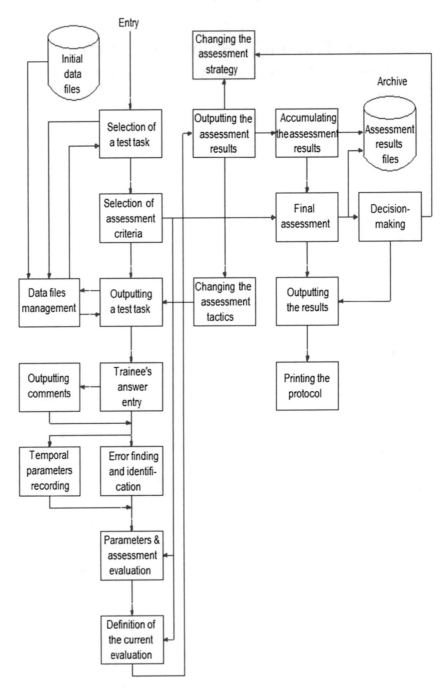

Fig. 8.1 Functional diagram of the knowledge and skills assessment process

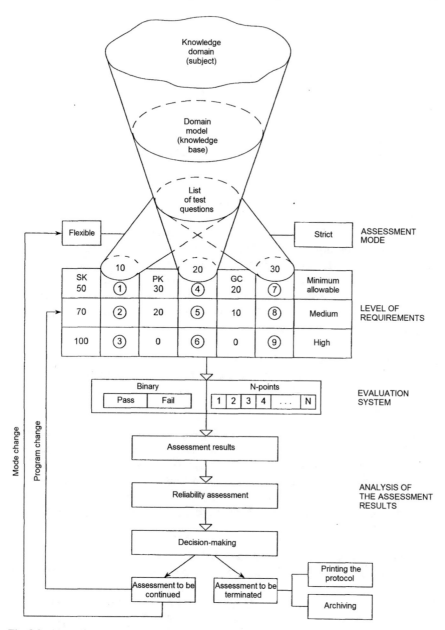

Fig. 8.2 Air traffic controllers' knowledge assessment using automated training aids

form of the final knowledge assessment protocol; archiving and printing the assessment results.

The automated training aids knowledge base is a set of logically combined and systematized portions of educational material connected by a single structure, consisting of test questions, reference answers, and comments on the trainees' actions. The knowledge base shall contain educational material that corresponds to the program of a particular discipline (subject) and meets the requirements of specialists' professional knowledge certification.

The basic operational modes of the automated ATC specialists training and assessment system are knowledge assessment, unsupervised work, section edition, help, archiving, and completion of work.

The mode is selected with a computer mouse. After that, the system switches to the appropriate operational mode.

Let us consider the knowledge assessment mode in more detail. This mode is basic and is designed to implement the functions of automated knowledge assessment. To do this, select the "Knowledge Assessment" function in the main menu of the system. Then, the controller to be trained (assessed) is registered, i.e., their personal and other data are entered. Next, the system is "tuned," i.e., the necessary parameters and assessment mode are selected. These are selected sequentially in the appropriate program menus using the mouse.

Assessment Parameters and Type Rating
Specify the type of *rating* for the controller to be tested: aerodrome controller; approach controller; radar-based approach final controller; Unified ATM System area controller.

Supervisory Unit A supervisory unit is selected, where the controller is authorized to perform their ATC functions. The *aerodrome controller* is authorized to work on at least one of the following ATC units: ASU, TSU, RSU, ARSU, ATC tower, local RSU, local control tower, local CC, and their combinations (in a prescribed manner); the *final controller—FSU* (LSSU); the *approach controller—ASU*, CSU, control tower; the UATM system *area controller—UATM* system ACs, UATM system remote ACs, local SU (LSU), RLSU.

Position The *position* of the tested specialist is selected: Auxiliary ATS personnel (assistant controller, informant controller, operator controller, etc.)—operator (O); controller (C); radar assistance controller; graphic (procedural) air traffic controller (GAC); aerodrome ATC instructor (AeroAI); area ATC instructor (AreaAI); senior SU controller (SSUC); senior aerodrome shift controller (SAeroSC); senior area shift controller (SAreaSC); aerodrome flight operations director (AeroFOD); area flight operations director (AreaFOD).

Assessment Type The appropriate *assessment type* is selected: Approval for unsupervised work; authorization for international flights ATS in English; renewal of the certificate; promotion; preparation for operations in the danger flight zone (DFZ); interruption in ATC operations (more than three months); by decision of the flight director (senior official); on the personal initiative of a specialist. The assessment type determines the assessment mode selected.

Assessment Mode The assessment mode has a set of relevant assessment parameters. These include the number of test questions (tests), the level of requirements, the presence (or absence) of feedback ("correct," "wrong"), prompts for the correct answers, the use of a timer, the time limit for preparing an answer (specific numerical values), the way of entering the answer, etc. Depending on the type and purpose of the knowledge test, the teacher (instructor) can select the necessary assessment mode: "flexible" or "strict." (see Fig. 8.3)

In the first case, a "flexible setting" of the system is possible, i.e., an arbitrary selection of assessment parameters, which makes it possible for the examiner to organize an assessment program at their own discretion. The "strict" mode has a fixed set of assessment parameters that define the appropriate assessment type, for example, "Assessment 1," "Assessment 2," "Assessment 3," etc. The strict mode is more preferable since it makes the trainees equal in testing conditions and implements the same assessment program and level of requirements. In addition, this mode is more convenient when organizing an archive, since it enables the presentation of the assessment results in a strictly defined form.

Level of Requirements

This parameter defines the requirements for the level of knowledge of ATC specialists. Each test question in the knowledge base of the ASKM has its own assessment criterion for the corresponding specialist category (controllers of certain supervisory units). As evaluation criteria, it is proposed to use solid knowledge (SK), proficient knowledge (PK), and general concepts (GC) (the description of these criteria is provided in the methodology for knowledge assessment.).

At the first stage, when assessing the knowledge of ATC specialists in the field of regulatory legal documents and special subjects (aeronautical meteorology, fundamental aerodynamics and aircraft performances, air navigation and navigational flight support, etc.), the above criteria are used when determining the requirements for the knowledge of the relevant test question for controllers having the appropriate *rating* (see the presentation form). The importance or significance of each of the questions included in the ASKM knowledge base is determined by their weighting factors in accordance with the evaluation criteria and is taken into account in the future when calculating the indicator of the trainee's level of training.

In assessing the controllers' knowledge of procedures, radio rules and phraseology, more stringent requirements apply for their level of training. The teacher (instructor) can select one of the three levels of requirements: minimum-allowable; medium; high—according to the assessment type.

The level of requirements is defined by the percentage ratio of the number of test questions with the corresponding evaluation criteria ("SK," "PK," and "GC").

The Number of Test Questions

The number of test questions (tests) in the test program can be arbitrary or fixed.

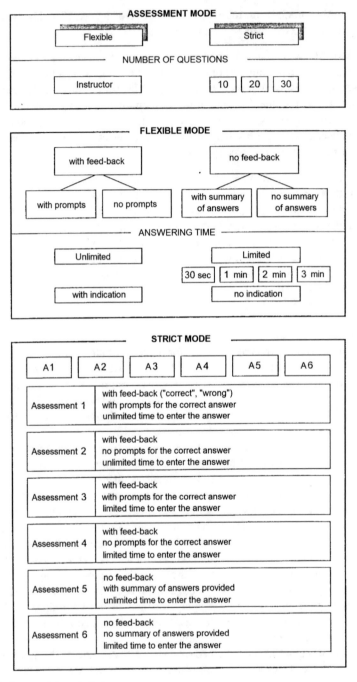

Fig. 8.3 Assessment mode features

When using the "flexible" control mode, the number of questions is set by the teacher (instructor) or is chosen by the trainee. The "strict" mode involves the selection of a certain number of test questions (10, 20, or 30).

Evaluation System

The instructor, depending on the assessment type and purpose, shall select the required system of knowledge evaluation: binary ("pass–fail"); or five-point grading system.

Examples of calculating the indicator of the trainee's level of training and the methods of grading (evaluation) are described in detail in the "Methodology for assessing the level of knowledge."

Assessment Section

Select the necessary section for knowledge assessment. The teacher (trainee) shall select a subject, topic or section from the database to be used in the assessment. In accordance with the "Guidelines for the Training of Civil Aviation Air Traffic Services Personnel," the ATSs databases include the following subjects: regulatory documents (Air Laws Regulations, Federal airspace regulations, Federal Aviation Regulations "Flights in the airspace of the Russian Federation," FAR "Planning and performing flights in civil aviation of the Russian Federation," table of reports on aircraft movement in the Russian Federation, etc.); work procedures, radio rules, and phraseology; aeronautical meteorology; fundamental aerodynamics and aircraft performances; air navigation and navigational flight support; radio engineering and lighting engineering support of flights and aeronautical telecommunications; English (for personnel performing ATS functions on international airways and in areas of international airports). The section "Work procedures, radio rules and phraseology" includes a list of test questions relating to flight rules, work procedures, radio phraseology for controllers who have appropriate ratings and approvals to work at certain supervisory units. The "English Language" section contains terminology, definitions, grammatical forms, procedures, radio rules, and phraseology in English, the knowledge of which is necessary for air traffic services on international airways and in areas of international airports. If necessary, or at the request of the system user, each of the test sections can be modified or supplemented with questions reflecting the specific work of controllers in a particular airport service. After selecting and entering the above parameters, the system switches to the knowledge assessment mode according to the established program.

Knowledge Assessment

In accordance with a given assessment mode, test questions of various types are randomly selected from the knowledge base and presented to the trainee.

The tested controller shall answer the questions selecting (forming) the correct answer from the suggested options and enter it into the system using a mouse or keyboard. For example, when using "select from the menu" answers, the trainee shall select the correct answer from the suggested alternative answers or enter the answer "yes," "no," or "I do not know"; when performing "direct answer input" tasks, enter specific numerical data, when solving problems of building an algorithm (sequence) of actions, select and consistently enter the necessary actions (operations) to form an answer. Available types of answers are contained in a special form of the ATSs test questions.

Output of the Assessment Results

At the end of the assessment cycle, the following "menu" is displayed on the computer screen: PRINT PROTOCOL; SUMMARY OF ANSWERS; MAIN MENU. This mode is designed to display the protocol of the assessment results on the screen or to print it for a particular ATC specialist tested by the relevant test program.

Summary of Answers

This mode is intended for reviewing test questions that were answered incorrectly. In this mode, the following information is displayed: the number of incorrect answers, the wording of the question, the trainee's answer, the correct answer, and the source (article or paragraph of the regulatory document or paragraph or page of the reference material used) where the question is considered.

Main Menu

This mode is intended to enter the main menu of the system.

8.2 RPMS Integrated Automated Training Aids Database Organization

The RPMS integrated automated training aids knowledge base has a hierarchical structure and includes the above knowledge domains (subjects). Below is the structure of the knowledge base for the "Unsupervised work" and "Knowledge Assessment" modes.

Structure of the RPMS integrated automated training aids knowledge base ("Unsupervised work" mode).

Whole Database. S 1. Regulatory Documents:

P 1. Air laws regulations of the Russian Federation (RF ALR);

P 2. Federal airspace regulations of the Russian Federation (RF FAirR);

P 3. Federal aviation regulations "Flights in the airspace of the Russian Federation" (FAR FARF);

P 4. FAR "Planning and performing flights in civil aviation of the Russian Federation";

P 5. Table of reports on aircraft movement in the Russian Federation (TS-95);

S 2. Controller Work Procedures, Radiotelephony (Radio) Rules and Phraseology:

P 1. Air traffic controller work procedures (CWP):

T 1. Methodical instructions;

T 2. Aerodrome controller work procedures: T 2.1 ACU CWP (Local ACU); T 2.2 TSU CWP; T 2.3 SSU CWP (Local SSU); T 2.4 ASSU CWP; T 2.5 Local ACT CWP; T 2.6 Local CSU CWP.

T 3. Final controller work procedures: T 3.1 FSU CWP (LSSU).

T 4. Technologies of work of approach controllers: T 4.1 CC (LSSU) CWP; T 4.2 ApC (GApC) CWP; T 4.3 Tower CWP.

T 5. Area controller work procedures: T 5.1 UATM System AC (Remote AC) CWP; T 5.2 LSU (RLSU) CWP.

T 6. Methods of air situation registration by the approach controller.

T 7. Symbols for the designation of the air and the meteorological situation in the control room traffic schedule.

T 8. Methods of air situation registration on the UATM AC (RAC) controller's tablet.

T 9. Table of the required minimum distance to perform a maneuver when crossing an occupied flight level (Appendix 4).

T 10. Calculation of the required minimum time interval at the point of entry into the area when the "Mach Number Method" is applied.

T 11. Methods of air situation registration by the LSU controller using a tablet.

P 2. Radio rules and phraseology during flight operations and ATC communications (RRP):

T 1. General provisions;

T 2. Terms and conventions;

T 3. General radio rules;

T 4. General radio phraseology;

T 5. Aircraft identification and separation and the use of secondary radars (SSRs);

T 6. Standard radio phraseology of air traffic controllers with aircraft crews: T 6.1 Taxiing supervisory unit; T 6.2 Runway supervisory unit; T 6.3 Circuit control; T 6.4 Landing (final) supervisory unit; T 6.5 Approach supervisory unit; T 6.6 ATM Area supervisory unit.

T 7. Radio communication rules in emergency and urgent situations.

T 8. Radio communication with vehicles and aerodrome facilities.

T 9. Exchange of information between controllers in the ATC process and radio phraseology.

S 3. Aeronautical Meteorology and Meteorological Support for ATC Units: P 1. Aeronautical meteorology; P 2. Meteorological support for ATS units.

S 4. Fundamental Aerodynamics and Aircraft Performances: P 1. Fundamental aerodynamics; P 2. Aerodynamics and flight dynamics; P 3. Aircraft performances.

S 5. Air Navigation and Navigational Flight Support: P 1. Air navigation; P 2. Air navigation flight support.

S 6. Radio Engineering, Lighting Engineering Support of Flights and Aeronautical Telecommunication: P 1. Radio engineering support of flights; P 2. Electric lighting support of flights; R 3. Aeronautical telecommunications.

S 7. English: P 1. Aviation English; P 2. Radio phraseology in ATC in English.

Structure of the RPMS integrated automated training aids knowledge base ("Knowledge assessment" mode)

S 1. Regulatory Documents

The database structure for this section is similar to the corresponding section of the "Unsupervised work" mode.

S 2. Controller Work Procedures, Radio Rules and Phraseology:

P 1. Rating—aerodrome controller:

T 1. Aerodrome supervisory unit (ASU, Local ASU);

T 2. Taxiing supervisory unit (TSU);

T 3. Runway supervisory unit (RSU, local RSU);

P 2. Rating—final controller:

T 1. Final supervisory unit (FSU);

R 3. Rating—approach controller:

T 1. Circuit control (CC, LSSU);

T 2. Approach supervisory unit (ASU, GASU);

P 4. Rating—area controller:

T 1. ATM area control center (AC (RAC));

T 2. Local supervisory unit (LSU, RLSU).

S 3. Aviation Meteorology and Meteorological Support for ATS Units

The database structure for this section is similar to the corresponding section of the "Unsupervised work" mode.

S 4. Fundamental aerodynamics and aircraft performances:

R 1. Fundamental aerodynamics;

P 2. Flight dynamics and aircraft performances.

For subjects S5, S6, S7, the database of the "Unsupervised work" mode is used.

8.3   Organization of the Knowledge Base for Subject "Controller Work Procedures and Radio Rules and Phraseology"

As an example, consider the organization of the knowledge base for subject "Controller work procedures and radio rules and phraseology." First of all, it is necessary to determine the requirements for air traffic controllers' level of knowledge. It is advisable to use the following evaluation criteria:

- *solid knowledge (SK)*—accurate (literal) knowledge and deep understanding of a clause, rule or article of regulatory documents regulating the controller's work (flight rules, work procedures, radio rules, and phraseology, etc.);
- *proficient knowledge (PK)*—the level of knowledge of special and applied subjects and work procedures that is necessary to solve ATC tasks correctly in each particular case;
- *general concepts (GC)*—the level of knowledge that provides a general understanding of the issues included in the list of special and applied subjects.

Each test question in the RPMS integrated automated training aids knowledge base shall have its own evaluation criterion for the corresponding specialist category (supervisory unit or workstation). For example, the analysis of tasks solved by air traffic controllers in various areas makes it possible to determine the relationship between the controllers' level of knowledge for the relevant supervisory units (see Table 8.1).

8.4   Organization of Remote Training for ATC Specialists

Remote training for ATC specialists is implemented on the basis of web technologies and provides the following features:

(1)   Multi-user mode. The web technology provides a centralized ATSs databases and knowledge bases on a secure server. All clients wishing to connect to the system pass an authorization procedure that is mandatory. The password system provides various access rights for various groups of ATSs users;
(2)   Centralization of databases that makes it possible to accumulate and store the results of ATSs users' training on the server, to form and output various reports on the training results;
(3)   The accessibility of the system via the Internet, which includes https encryption. At the same time, it is possible to organize access for ATC specialists both to training materials and to training results;
(4)   Embeddability. If necessary, it is possible to embed web pages into any other ATSs systems using the standard "web browser" component.

The system is based on the client–server principle, thus making it possible to: implement a centralized database with a multi-user mode; access the knowledge base/database and training results via the Internet, including encryption; use other technologies to extend the functionality (Flash, Java, Silverlight, etc.) or create and use own client application if the functionality of the standard browser is not sufficient.

Server part   The knowledge base/database is located on the server and managed by the administrator. The administrator monitors the health of the knowledge base/database, edits it, and distributes users into groups with different access rights.

Client part   All clients wishing to connect to the system pass the mandatory authorization procedure. Depending on the group of users, the client can access various training materials, tests, reports on training results and fill in/edit the knowledge base/database.

Section Review

1.   Main tasks and method of organizing the automated training system for air traffic control specialists.
2.   Block diagram of the air traffic controllers' knowledge assessment process in the automated training system.

Table 8.1 Characteristics of dispatcher qualifications

Supervisory unit \ Rating	Aerodrome controller			Final controller	Approach controller			Area controller		
	ASU (Local ASU)	TSU	RSU (Local RSU)	FSU (LSSU)	CC (LSSU)	ASU (GASU) Radar	ASU (GASU) GC	UATM AC (RAC) Radar	UATM AC (RAC) GC	LSU (RLSU)
Aerodrome controller — ASU (Local ASU)	1									
Aerodrome controller — TSU	1	2								
Aerodrome controller — RSU (Local RSU)		2	3							
Final controller — FSU (LSSU)		2	3	4	5					
Final controller — CC (LSSU)			3	4	5					
Approach controller — ASU (GASU) Radar			3	4	5	6	7			
Approach controller — ASU (GASU) GC					5	6	7			
Area controller — UATM AC (RAC) Radar						6	7	8	9	10
Area controller — UATM AC (RAC) GC						6	7	8	9	10
Area controller — LSU (RLSU)								8	9	10

Knowledge evaluation criteria:

■ - solid knowledge (SK)   ▨ - proficient knowledge (PK)   ▦ - general concepts (GC)

3. Operational modes of the automated training system for ATC specialists and their features.
4. Organization of the database of the automated training system for air traffic controllers.
5. Organization of remote training using the automated system for training air traffic controllers.

Chapter 9
ISS Special Software

ISS special software (SSW) can be divided into two main complexes: training sub-system SSW; modeling and evaluation subsystem SSW.

The principle of construction of the training subsystem is based on the use of hardware and software of the simulated automated air traffic control system. This ensures the full adequacy of the implementation of air traffic control processes in the ISS. The modeling and evaluation subsystem SSW is intended to solve the following main tasks:

- configuring workstations for several training modules;
- maintaining the same exercise time for all facilities of one training module;
- simulation of the movement of various types of aircraft in accordance with their tactical, technical, and aerodynamic performances;
- simulation of special cases and emergency situations;
- generating data of multi-radar information processing of simulated radars and transmitting it to the RPIPS;
- simulation of information on the bearing;
- simulation of meteorological data and the issuance of this information to the AWS-C instruments;
- generation, amendment and transmission of planning information to the trained controller;
- implementation of the air situation window on the OP (instructor) workstation software with an interface similar to the interface of the trained controller;
- recording, replaying, and restoring the exercise progress synchronously with the OP and the trained controller's speech information;
- registration of the ATC parameters providing the means of recording the con-trollers' actions during the training process;
- ensuring the operation of the modeling and evaluation SSW in real and fast-forward time with possible short stops;
- preparing any given airspace structure and creating, correcting, and storing the planning information used in the exercise;

© Springer Nature Singapore Pte Ltd. 2020
Bestugin A.R. et al., *Air Traffic Control Automated Systems*,
Springer Aerospace Technology, https://doi.org/10.1007/978-981-13-9386-0_9

– implementation and storage of the library for exercises, backup, and recovery capabilities.

The input to the script process is as follows:

– information base of exercises on the basis of the training scenario;
– commands entered from OPs (instructors)' consoles;
– messages received over the local network from AWS-C, the training supervisor and the system administrator.

The information base of exercises refers to the collection of data on the structure of the airspace and ground tactical situation, scheduled tasks for the simulated aircraft, meteorological conditions, aircraft parameters, radar parameters, and elements of the graphic information. Each exercise has its own name, and an exercise library is created from the pre-defined exercises.

The modeling and evaluation subsystem SSW includes: air and ground situation models; models of radio and lighting means of flight operations support; a planning information simulator; management and configuration software; documentation and replay package; a set of training exercises; and an automated assessment package.

The air situation model contains: a model of the aircraft movement; an emergency model; a model of weather conditions; software package for automatic formation of trajectories for moving objects on the airfield; model of the visual aerodrome environment; and radio-technical equipment (RTE) failure model.

The exercise preparation package contains: a zone structure creation block; an exercise formation block; a creation and adjustment block for flight plans.

The automated assessment package includes the software for the following: collecting and evaluating information in the training process; aircraft conflict definitions; processing and documenting the exercise results.

9.1 Aircraft Movement Model

The movement of various types of aircraft, both on the ground and in the airspace is modeled in accordance with their tactical, technical, and aerodynamic performances in real time. When simulating the aircraft movement, the effect of the wind is taken into account. The range of changes in aerodynamic performances is set based on actual flight conditions. It is advisable to simulate the aircraft movement in the ISS by imposing two methods: a method for modeling movement along specified trajectories through given points and a method for movement modeling based on dynamic Euler equations. In addition, the trajectory of the aircraft movement in the ISS is a function of the altitude-heading commands sent by the controller. The aircraft movement along specified trajectories is built using points. The trajectory of the aircraft movement is represented by a set of points. A point is a certain given condition of the aircraft movement; upon fulfillment of this condition, a transition to the next point occurs, which is selected by specially defined rules. With this set of points, one can specify any trajectory.

The position of the aircraft is determined at discrete instants of time. The coordinates of the aircraft are calculated not only discretely—once per cycle of calculation, but also after the event has occurred; the time interval is a floating value. An event is the achievement of any given parameter (speed, altitude, heading, time, etc.) obtained as a result of the controller's control command entered from the instructor's console, or reaching a certain point.

The model of the aircraft movement between points or after the controller's command is built on the basis of the aircraft aerodynamic performances. The model is based on the system of dynamic Euler equations in a related coordinate system. The moments acting on the aircraft are assumed to be balanced by the corresponding deviations of the controls. It is considered that all the forces acting on the aircraft are applied to its center of mass.

The movement of an aircraft under the action of aerodynamic and thrust forces in a related coordinate system is described by a system of Eqs. (9.1). The engine thrust is directed along the longitudinal axis of the aircraft, and the sideslip angle is small and can be neglected. All calculations are performed in the Cartesian coordinate system, relative to some conditional point; the X-axis is directed to the north.

$$
\begin{aligned}
\frac{dv}{dt} &= g \cdot \frac{P \cdot \cos(\alpha) - X}{Q - \sin(\theta)} \\
\frac{d\theta}{dt} &= \frac{g}{V} \cdot \frac{Y \cdot \cos(\gamma) + P \cdot \sin(\alpha)}{Q - \cos(\theta)} \\
\frac{d\varphi}{dt} &= -\frac{g}{V \cdot \cos(\theta)} \cdot \frac{Y}{Q \cdot \sin(\gamma)},
\end{aligned}
\tag{9.1}
$$

where g is the downward acceleration (acceleration of gravity); θ is the angle of trajectory inclination in the vertical plane; γ is the roll angle; φ is the heading; V is the true speed; P is the aircraft thrust; Y is the lifting force; X is the drag force; Q is the force of gravity (aircraft weight); α is the attack angle.

All forces are considered to be applied to the aircraft center of mass. The range of selected aerodynamic performances is determined by using graphical dependencies of aerodynamic performances for each given type of aircraft and flight mode; thus, to take into account the drag force, a graphic dependence on the flight speed and aircraft type is used for the zero lifting force.

The speeds are selected from the range of operational speeds and altitudes of a given type of aircraft. The graphs of the lifting force coefficient versus the induced drag coefficient at various flight speeds are used.

Also, graphic dependences of the allowable value of the lifting force coefficient on the speed are used.

The system (9.1) is the main one for calculating the parameters of the aircraft movement, and to calculate the aircraft coordinates, the system of kinematic Eqs. (9.2–9.4) is used.

$$
\frac{dx}{dt} = V \cdot \cos(\theta) \cdot \cos(\varphi)
$$

$$\frac{dy}{dt} = \frac{dh}{dt} = V \cdot \sin(\theta)$$

$$\frac{dz}{dt} = -V \cdot \cos(\theta) \cdot \sin(\varphi) \tag{9.2}$$

$$v[i+1] = v[i] + \frac{dv[i+1]}{dt} \cdot \Delta t$$

$$\varphi[i+1] = \varphi[i] + \frac{d\varphi[i+1]}{dt} \cdot \Delta t$$

$$\theta[i+1] = \theta[i] + \frac{d\theta[i+1]}{dt} \cdot \Delta t \tag{9.3}$$

$$x[i+1] = x[i] + \frac{dx}{dt} \cdot \Delta t$$

$$z[i+1] = z[i] + \frac{dz}{dt} \cdot \Delta t$$

$$h[i+1] = h[i] + \frac{dy}{dt} \cdot \Delta t \tag{9.4}$$

The systems of Eqs. (9.1–9.4) fully describe the aircraft movement in space, make it possible to calculate all the parameters of its movement and their change over time, if the engine operational mode and thrust and aerodynamic performances of a given aircraft type are set.

The value of ground speed and ground angle is determined by the following formulas (9.5–9.7).

$$W = \sqrt{V^2 + U^2 + 2 \cdot V \cdot U \cdot \cos(\psi)} \tag{9.5}$$

$$\varphi_c = \arcsin\left(\frac{U \cdot \sin(\psi)}{W^2}\right) \tag{9.6}$$

$$\varphi^* = \varphi + \varphi_c, \tag{9.7}$$

where φ is the initial heading; V is the true speed; U is the wind speed; W is the ground speed; ψ is the heading angle between the V and U vectors; φ_c is the crab angle; and φ^* is the ground heading.

The advantages of the presented method of modeling the aircraft movement in training systems are that it allows significant increasing in the similarity of the simulated flight to a real one. It takes into account the relationship of longitudinal and transverse movement, the dependence of speed on the altitude, weight, and maneuvers performed by the aircraft. The simulation of the movement dynamics this method takes into account a wide range of changes in the speeds and altitudes of modern types of aircraft, as well as performances of the aircraft under the engine limit conditions. External control actions are as follows: the exercise of the controller's commands is performed by inputting control commands from the AWS-OP functional keyboard.

In order to control the aircraft, the following basic functions are implemented by the controller's commands when modeling the ground and air situations:

- taxiing and stopping the aircraft moving on the aerodrome maneuvering area;
- control of arriving and departing aircraft;
- setting an altitude (flight level) both in meters and feet;
- setting the vertical rate of climb/descent;
- setting an altitude (flight level) to a specified point;
- setting parallel displacement;
- setting a heading;
- setting a turn at a certain angle;
- setting a bank angle when turning;
- setting the direction of the turn;
- setting flight speeds, both in km/h and in knots;
- setting the Mach number;
- disabling/enabling "A" and "C" modes of the transponder;
- setting a new code of the transponder;
- simulation of the operational modes of the transponder;
- flying to a specified point;
- flying to the holding area and staying within the holding area;
- changing the flight route;
- dispatching aircraft following SID/STAR trajectories;
- commands that simplify or complicate the air situation (changes in weather conditions, introduction of additional aircraft, and introduction of special cases).

9.2 Automatic Generation of Movement Trajectories for Vehicles on the Airfield

The current coordinates of the aircraft position on the airfield are based on the use of the kinematic equations of motion for its center of mass. The coordinates are calculated in accordance with a given calculation cycle or according to events (when the controller gives commands or when the parameters set for a given point are achieved). The aircraft movement along the specified trajectories presents point-to-point movement, and the trajectory is a sequence of points. A point is a certain condition of the aircraft movement, after which the transition to the next point occurs. The task of finding the shortest path from graph theory is used to automatically generate the trajectories of vehicles moving on the airfield. To do this, at the stage of preparation of exercises, a database of taxiways, aprons, and parking places is created. Data on taxiways consist of a set of points in the Cartesian coordinate system with a corresponding report, if a radio exchange with the start/taxiing controller is required at this point, and with a radius, if a turn of a certain radius is required at this point. Aprons are parking places with unique names and the same taxi routes. They indicate taxi routes in the form of taxiway names for taking off and landing regarding each of the runways at the airport. A parking place (PP) is a coordinate with the heading belonging to the aerodrome and apron indicated in it, and indicating the exit and entry points (with or without a towbar). The taxiways form a closed graph, which creates a matrix of connections. The method of automatic formation of movement trajectories for vehicles on the airfield consists in selecting the shortest path in this graph along the specified route: from the PP to the take-off point or from the landing point to the PP, along the route indicated for the apron on which the PP is located; from the current position of the vehicle to the specified PP when landing or to the take-off point during takeoffs if taxiing commands are given.

The taxiing command indicates the taxiways to be used to drive along, an apron, a PP and a free PP, if necessary, through which one can enter or leave the PP. Thus, the information support of the automatic generation of the movement trajectories for vehicles on the airfield is a variable set of matrices and graphs, the quantitative composition and completion of which are determined and completed when the model is adapted to a specific aerodrome.

9.3 Simulation of Radio Technical Facilities for Flight Operations Support

In the ISS, models of radiotechnical facilities for flight operations support (RTF FS) provide for the simulation of information from the following information channels: primary surveillance radar; secondary surveillance radar; landing radar; airfield surveillance radar; automatic direction finder; IRD information; ADS-B and ADS-C information; and CPDLC data transmission links.

Since the software package for modeling the air situation with a given calculation cycle constantly provides coordinate information and the results of secondary, tertiary and multi-radar processing of information, it does not make sense to simulate analog signals of the RTF FS in the ATC simulators. The task of simulating the functioning of information channels (Chi) in ATC simulators is reduced to imposing resulting errors [1] corresponding to each simulated channel (δf(Chi)) on the information (IMP) simulated by the air situation modeling package. The structure of the RTF FS modeling organization is presented in Fig. 9.1. Considering the high level of information processing at the output of various RTF FS packages and in ATC control systems themselves, the task is reduced to modeling the mean values of errors with a normal (Gaussian) distribution law and zero expectation. Mathematical models for an omnirange radar shall also take into account such characteristics as visibility areas and radiation patterns. The radar information is formed taking into account the radar rate of surveillance.

The input information on all radars is located in the exercise database. Based on the input information from the air situation modeling package, the probability of detecting a target in the radar visual field can be calculated. The radar visual areas of

Fig. 9.1 Structure of the RTF FS modeling organization

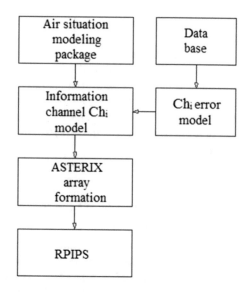

the primary and secondary channels are set by the piecewise linear approximation method taking into account the locations of the radars. If the aircraft is in the radar visual field (in its sight), then the range of confident detection is determined, which is given by the intersection points of the aircraft azimuth and one of the lines that approximate the line of sight. If the current range at the corresponding azimuth does not correspond to the confident detection range, then the detection probability is zero. Directly in the area of confident detection, the probability of detection shall be calculated from the signal-to-noise ratio. Considering the high level of signal processing in modern RTF FS and information processing in modern automated ATC systems, the probability of detection in the confident detection area is usually taken as 1 (one). False targets can be simulated as follows.

The number of false targets for the exercise period is a random value with an upper limit (depends on the type of equipment and is set in the database). The coordinates of false targets are generated as random variables with a uniform distribution in the radar visual field. In accordance with the rate of surveillance of the simulated radars, an array of radar information is generated regarding all aircrafts within the visual fields of all locators installed in the simulated space. The generated array is transmitted over the LAN to the radar and planning information processing server (RPIPS). Arrays of radar information are usually of the ASTERIX format. To increase the reliability, it is possible to operate the RPIPS simultaneously in the system mode and in the BY-PASS mode, i.e. an additional array of radar information is generated in the same ASTERIX format. The information on the air situation obtained from ADS-B, ADS-C, ADF and CPDLC is modeled and generated in a similar way: by imposing errors with probabilistic characteristics inherent in the type of airborne and ground equipment on the true information obtained from the air situation modeling package. At the output, an array in the universal ASTERIX format is generated. Information is given to the RPIPS at a speed that corresponds to the functioning of the simulated real navigation aids. A direction finder is simulated by transmitting information on the detected aircraft at the time of receiving a one-time command of direction finding from AWS-CD or in automatic mode during a radio communication of a trained controller with the operator-pilot playing for the detected aircraft. The information is transmitted to the RPIPS over the LAN in the ASTERIX format. In this case, the AWS-C displays the bearing line in accordance with the received information. At the end of radio transmission, the bearing line is erased. When calling another aircraft, the bearing line changes.

9.4   Organization of RTF FS Faults Simulation in the ISS

Testing the skills of air traffic controllers in emergency situations: in case of failure of some RTF FSs, the ISS implements the functionality of simulating failures of one or several information channels simultaneously simulated by the TS. A failure is simulated in the following sequence. On the information field of the AWS-TS screen, an "RTF FS fault" window is provided with an extension table for all information channels modeled in the ISS. The training supervisor can specify a set of faults in any combination. A fault means prohibition on the issuance of information for given aids. After "recovering," information starts to come in again. So, when modeling the failure of the radar secondary channel, only information on aircraft coming through the primary channel is transmitted via the LAN to the RPIPS, and the tagging form (tag) is no longer displayed on the AWS-C ASD. When the "Recovery" command is sent, the issue of information over the specified channel is activated and the tags are restored. When a failure of an ADS receiver is simulated over the LAN, the transmission of information that simulates an ADS channel stops. If a radio direction finder fails, information on the bearing is not transmitted to the AWS-C. Changes in the display of information on the AWS-C ADS screens when certain commands of the RTF FS are sent depend on the functionality of a specific ATC control system, the software of which is installed in the ISS. When simulating failures of radio channels, it is prohibited in the ISS to issue information on the bearing. Partial or total radio communication failure can be simulated. In the event of a partial radio communication failure, the aircraft continues to move along the predetermined trajectory and follows the air traffic controller's instructions, but does not report on the accomplishment of these instructions; the corresponding reports are not displayed on the AWS-OP screen. On the AWS-C ADS screen a flag of the database appears in the tag data. In case of a complete failure of the radio communication, the aircraft continues to move along a predetermined trajectory, does not follow the air traffic controller's instructions and does not report on the accomplishment of his/her actions. The corresponding reports are not displayed on the ARS-OP screen. Aircraft taking off within the circle center return to their airfield. On the ARM-S ADS screen, a flag of the database appears in the tag.

9.5 Simulation of Meteorological Conditions

Meteorological conditions in the ISS are simulated to solve the following tasks: considering the meteorological conditions in the air and ground situation models specified in the exercise; organization of issuance of data on weather conditions in the controller's area of responsibility to the AWS-C in the process of exercising.

The meteorological conditions for each exercise are set in the database for each control area in the form of the "Meteodata" table. The formation of a database of weather conditions using a tabular window technology is shown in Fig. 9.2.

The data from the "Meteodata" table of elements describe the meteorological situation of the selected area. In the "Meteodata" table, the parameters characterizing each altitude layer are set: the upper boundary of the layer, wind speed, meteorological wind direction, meteorological phenomena; meteorological conditions on the runway: temperature, pressure, visibility on the runway, strip coating, friction coefficient on the runway, and humidity. The wind characteristic is set: direction, speed by layers, and weather phenomena.

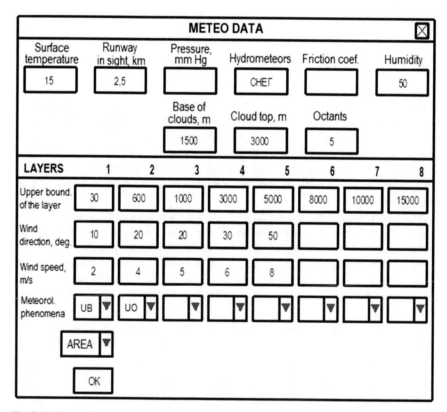

Fig. 9.2 Generation of the meteorological information database

In models of air and ground situation, when modeling the aircraft movement, the aircraft speed vector takes into account the wind speed vector component.

Wind speed and direction change when the aircraft moves from one altitude layer to another. The model of the visual situation in the aerodrome area also takes into account the wind speed, visibility range, and the base of clouds on the runway; commands of simulation of snow, rain, and fog can be given. These data are also initially set when preparing the exercise by setting the initial data in the "Meteorological data" table. Clutters from the clouds are simulated, with the possible setting of clouds during the preparation of the exercise and the rapid setting and removal of clouds with an indication of the location. The movement of clouds is simulated in accordance with the wind direction and speed. The model of the air situation provides for the practicing the trained controller's command for avoiding a thunderstorm with the subsequent return of the aircraft to the set flight trajectory.

In the course of training, one can quickly change the meteorological data. When changing meteorological data, operational data are entered into the "Meteorological data" table, which will be used in the airspace modeling software package and transmitted over the LAN to AWS-C at a given speed for displaying weather data on the ADS screen.

9.6   Emergency Simulation

The possibility of practicing the skills and abilities of air traffic controllers in the event of special cases and emergencies during the flight is a unique ISS functional characteristic. The simulation of emergency situations or special cases can be initialized as follows: when creating an exercise in the preparatory phase, with each special case implemented after some time specified in the flight plan; when practicing a special control command sent in the process of the exercise as decided by the training supervisor.

When exercising a special case, an alert message from the aircraft crew is displayed in the AS line for the controller. On the AWS-C ASD screen, a flag of the database is displayed in the tag as a signal informing of the need for assistance.

The following main simulated messages on special cases and emergencies are displayed for practicing emergency situations on the AWS-C ASD: on-board fire; engine failure; cabin decompression; loss of radio communication; aircraft hijacking; and disorientation.

Here, it is possible to cancel a special case, and in some cases, the continuation of the aircraft flight. Consider the principles of simulating the listed messages.

On-Board Fire   It is possible to simulate on-board fire on the ground and in the air. In this case, the appearance of a database flag is simulated on the ASD in the tag. The AWS-OP displays a special case report. When the aircraft fires before reaching the lift-off speed, in the aircraft movement model, the aircraft stops the run, decelerates, turns off the engines, and initiates the command to simulate a fire vehicle call; after extinguishing the fire, it simulates the aircraft towing to a given parking place.

If a fire occurs during the takeoff, the aircraft has lifted off from the runway and is climbing, a turn is modeled for a heading opposite to the landing one, and landing on the runway is simulated. The aircraft brakes, turns off the engines, and waits for the fire vehicle; after extinguishing the fire, the aircraft is towed to a given parking place. In case of fire in the circling area, a maneuver is simulated to stop the flight along the route followed by an approach to the nearest airfield. When the aircraft fires in areas after the circling area, the aircraft simulates an emergency descent to the circling height, the fuel is drained and then the landing is simulated, depending on the situation: either on the nearest airfield or outside the airfield. In case of fire on-board an arriving aircraft, it lands on the runway and decelerates, turns off the engines and waits for the fire vehicle. After extinguishing the fire, the aircraft towing to a given parking place is simulated.

Engine Failure In case of failure of all engines, the gliding distance is calculated at the current altitude of the aircraft using the following formula:

$$L_{GL} = C[(H_{curr} - H_{end}) + (V_{curr}V_{curr} - V_{end}V_{end})/2g] \qquad (9.8)$$

where C is the coefficient of aerodynamic quality of the aircraft; H_{curr} and V_{curr} are height and velocity at the start of gliding; and H_{end} and V_{end} are height and velocity at the end of gliding.

An aircraft landing at the nearest airfield is simulated, if the calculated gliding distance can ensure landing of the aircraft, otherwise landing at any suitable site is performed.

If the aircraft simulates a failure of fewer than all engines, then landing is either at the destination aerodrome or at the alternate aerodrome, depending on the aircraft position. When simulating (an) engine failure(s) at the airfield during takeoff or landing, the aircraft is towed to a given parking place. The AWS-C ASD displays a database flag in the tag. The AWS-OP displays reports on the failed engines indicated and the aircraft landing.

Cabin Decompression The AWS-C ASD displays a database flag in the tag. The AWS-OP displays a special case report. In case of cabin decompression, depending on the flight height, the following actions are simulated:

(a) if the current value of the flight height does not exceed the value allowable for human physiology, the aircraft continues to move along the given route. The AWS-C ASD displays a database flag in the tag;

(b) if the current value of the flight height is more than the allowable value, the aircraft departs from the assigned track, performs an emergency descent to the value of the height physiologically acceptable and continues the flight along the assigned route. The ASD displays a database flag in the tag.

Complete Failure of the Radio Communication When simulating a complete failure of the radio communication, the aircraft movement along a previously defined trajectory is modeled, and the aircraft does not follow the air traffic controller's instructions and does not report on its actions. No reports are displayed on the AWS-OP screen. Simulated departing aircraft located in the circling area perform a maneuver to return to their airfield. The ASD displays a database flag in the tag. The AWS-OP displays a special case report.

Aircraft Hijacking In case an aircraft hijacking is simulated, the movement along a given route is simulated. The AWS-C ASD displays a database flag in the tag. The AWS-OP displays a special case report.

Disorientation When simulating disorientation, the aircraft deviates from the flight path and continues to move along a parallel path until it receives instructions from the air traffic controller. The ASD displays a database flag in the tag. The AWS-OP displays a special case report.

When simulating emergency situations at the aerodrome, the following incidents are simulated: aircraft landing with the landing gear retracted; destruction of tires, gear breakage during takeoff and landing; on-board fire and smoke; aircraft runoff outside the airfield during takeoff, landing, and taxiing; collision of an aircraft with the ground or ground obstacles; collision between an aircraft and surface transport.

Aircraft Landing with the Landing Gear Retracted On the runway, when an aircraft landing with the landing gear retracted is simulated, the aircraft lands on the fuselage, decelerates and, in case of fire, the subsequent actions are similar to those described below in the "On-board fire and smoke" part. The radar displays a database flag in the tag. The AWS-OP displays a special case report.

Destruction of Tires, Gear Breakage During TakeOff and Landing In case a gear breakage is simulated, the aircraft decelerates to a stop, turns off the engines, and waits for the transport to tow it to a parking place. The radar displays a database flag in the tag. The AWS-OP displays a special case report. The visual part of the simulation scenarios of these two cases will look as follows: when simulating an aircraft landing with the landing gear retracted, destruction of tires and gear breakage, the shape, location, and color of the corresponding aircraft structural elements (landing gear, tires, and lower part of the fuselage) change.

Onboard Fire and Smoke When an onboard fire on the runway is simulated, the aircraft stops the run before reaching the lift-off speed, stops, turns off the engines, and waits for the fire vehicle; after the fire is extinguished, the aircraft is towed to a given parking place. When a fire on-board an arriving aircraft is simulated, it lands on the runway, where it stops, turns off the engines, and waits for the fire vehicle; after the fire is extinguished, the aircraft is towed to a given parking place. The AWS-C ASD displays a database flag in the tag. The AWS-OP displays a special case report. In this case, the visual part of the scenario will look as follows. In case of onboard fire and smoke, a visual model of fire and smoke is created in the area of the aircraft body where the engines are located. The surfaces of the aircraft in the fire area are darkened.

Aircraft Runoff Outside the Airfield During Takeoff, Landing and Taxiing
When an aircraft runoff outside the airfield is simulated, the aircraft stops, turns off the engines and waits for the transport to tow it to a parking place. The AWS-OP screen displays an accident report, and a distress flag lights up. When an aircraft runoff outside the airfield during takeoff or landing is simulated, the aircraft decelerates, stops, and turns off the engines. If there is no fire, it waits for the transport to tow it. In case of fire, the actions are similar to the case of "On-board fire and smoke." The AWS-OP screen displays an accident report, and a distress flag lights up. The model of the visual part will look as follows. When an aircraft runoff outside the airfield is simulated, an image of a banked aircraft with wheels stuck in the ground is displayed.

Collision of an Aircraft with the Ground or Ground Obstacles If the landing mode is incorrectly selected, it may cause a collision of the aircraft with the ground or ground obstacles. When a collision with the ground or ground obstacles is simulated, the aircraft breaks up. The ASD stops displaying the tag data and the aircraft symbol.

Collision Between an Aircraft and Surface Transport When a collision with surface transport is simulated, the aircraft stops moving, turns off the engines, and waits for the transport to tow it to a parking place. The AWS-OP screen displays an accident report, and a distress flag lights up. The visual part of the scenario for these two cases will look as follows. When an aircraft collides with the ground, other aircraft, special vehicles, and a model of a destroyed aircraft or vehicle is displayed with an image of construction fragments scattered on the ground or the airfield.

9.7   Documentation and Replay Package

The documentation and replay package for training exercises is designed to recreate situations that occurred during the exercise, to control the trainees' actions and to analyze and correct their mistakes. The documentation and replay package records the training process, including dynamic air situation (radar and navigation information), meteorological information, planning information, voice information, the state of the current information display windows, and command texts with all parameters entered from the service personnel AWSs (AWS-OP and ARM-TS).

The required mode of operation of the documentation and replay package is selected and adjusted in the Run and Configuration dialog box of the training system at the preparatory phase. In the process of starting the training, one of the training modes is selected: "No recording," "Recording." The "No recording" replay mode is used for debugging and verification of a newly created exercise. In this mode, no records are made. In this mode, it is possible to accelerate the time scale by a specified number. When selecting the "Recording" mode, the documentation tools will ensure that all necessary information is recorded on the hard disk for subsequent replay of the exercise progress. The record of each exercise is identified by the name of the current scenario, the date and time of the exercise. At the time of starting, the user can also enter any symbolic name to further identify the current record. The recorded information includes a list of parameters necessary for the replay of the exercise: parameters of all simulated aircraft; current meteorological information; current planning information; commands sent from the AWS-OP and AWS-TS; and voice information of simulated radio communication channels.

In the "Recording" mode, a file is created containing a list of the names of the recorded exercises and the time they were created. In this mode, an accelerated time scale is not possible. If the "Replay" mode is selected in the Run and Configuration dialog box, the documentation and replay package makes it possible to find the necessary training process and then replay it. The following replay modes are provided: simultaneous replay of all recorded information in real time; fast replay; and fast recovery of the module state at a specified time followed by replaying the exercise.

From the methodological point of view, the user shall be provided with the capability to consequently replay selected fragments of the training exercise, setting the start time and stopping the replay at the right time, and skip intermediate time intervals. When the module is in the "Replay" mode, the user can pause the replay, continue it, change the time scale, and stop the replay. From the "Replay" mode, it is possible to switch to the training mode ("No recording", and "Recording"). This allows consideration of several various options for the development of a particular situation that has evolved at some time point in the airspace. In addition, this feature allows quick recovery of the module in the event of a failure, and continuation of the training from the moment when the failure occurred.

9.8 Simulation of the Visual Airfield Environment

The model of the visual airfield environment (MVAE) ensures the creation of the visual and sound environment on the Tower ARM-C that is adequate to real external environment of the Tower controllers in the process of aircraft crew management and aerodrome special equipment management. The MVAE shall reproduce the audio-visual sensations that ensure the full and correct transfer of the actual flight control process to the appropriate Tower AWS-C and help form the necessary professional skills.

When reproducing the virtual scene MVAE observed with the Tower AWS-C, the following similarity conditions shall be observed: spatial similarity determined by the adequacy of the geometric dimensions of the scene elements, their color, and the texture of the materials; dynamic similarity ensured by the authenticity of the illusion of moving objects' movement (airplanes, helicopters, towing vehicles, birds, etc.);

adequacy of the characteristics of the environment in which the objects of the scene are located—illumination, brightness, contrast, chromaticity. and transparency.

The structure of the visual airfield environment simulation is presented in Fig. 9.3.

The most important MVAE function is the generation of a three-dimensional image of the real situation, taking into account the observer's position. The MVAE simulates the airfield environment, including the following: runways, taxiways, aircraft parking areas, buildings, high intensity lights, and lighting equipment, as well as mobile objects: airplanes, tugs, and towing vehicles.

The airfield environment visually observed from the Tower workstation tends to be constantly changing. Firstly, due to changes in the angle of observation (turning the head, using binoculars). Secondly, due to changes in the position of aircraft and surface vehicles. To track the dynamics of this changing environment, its model is recalculated with a certain frequency, ensuring real-time operation. At each cycle of the calculation, a projection of a three-dimensional situation model is created on the screen, and a frame is generated that is displayed on the screen. The MVAE functions in the local area network of the package and exchanges information with the ASGS. In order to ensure the ASGS MVAE functioning, the following is provided:

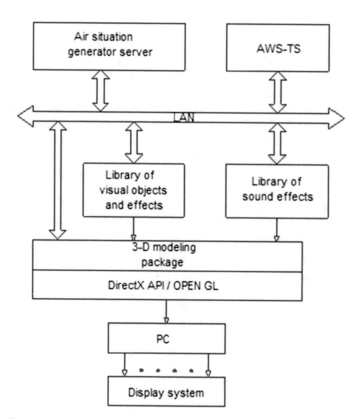

Fig. 9.3 Structure of visual airfield environment simulation

- coordinates for all aircraft and surface mobile objects are calculated and transmitted to the MVAE;
- messages are generated through the AWS-OP or AWS-TS that initiate changes in the visual environment on the trainee's console;
- commands of the training supervisor are generated and transmitted to the MVAE that complicate the conditions of observation for the trainee by introducing clouds, fog, rain, snow, or a flock of birds along the predetermined path into the scene;
- trained Tower controllers' commands are transmitted to the MVAE through the AWS-OP to control surface vehicles and aircraft after landing or before takeoff. The MVAE functions include simulation of the following characteristics of the air environment: sky; scene illumination; cloudiness; fog of variable intensity; and precipitation in the form of snow and rain.

The characteristics of the environment depend on the following parameters: time of day (determines the level of light and the color of the sky); wind speed (determines the speed of the clouds); the height of the cloud base (determines the visual position of the cloud on the display device); and fog intensity (changes visibility range).

These parameters are set in the description of the initial state of the visual scene, but can be changed on the AWS-TS and transmitted over the LAN to the MVAE to change the characteristics of the visual environment for the training. The MVAE provides simulation of mobile objects: airplanes, helicopters, special vehicles (tugs, towing vehicles, and tankers), and flocks of birds. Three-dimensional models of visual images of objects take into account the geometric dimensions, the signal equipment, shape and color characteristic of all specified types of aircraft, and special vehicles. The trainee can control the image on the screen through manipulations with an imaginary camera pointed at the airfield from the observer's position. The camera can be rotated around the vertical and horizontal axes using the computer mouse. When positioning the mouse cursor to any point of the scene and double-clicking the left mouse button, one can zoom in on the image, that is, simulate the "binocular mode."

The MVAE operates on a PC and is implemented as an application in the Windows XP OS (or another operating system with DirectX or OPEN GL features) in C++ and uses standard OS tools:

- basic OS functions: memory and external devices management, multitasking, queues and window messages servicing;
- graphical user interface supported by gdi32.lib, comdlg.lib, user.lib libraries.
- DirectX multimedia package providing the operation of two-and three-dimensional graphics and data transmission over a local network.

It is advisable to use the ninth version of DirectX in the MVAE, which has improved performance compared with previous versions. It is also advisable to use additional features of the hardware performance for graphic operations using shaders—that is, fragments of the program text written in a special language like an assembler language. This package provides access to two programmable pipelines embedded in the graphics processor of the video adapter for processing vertex and pixel shaders.

This significantly improves the MVAE performance and expands the possibilities of saturating the visual scene with additional elements. When creating images of surfaces, it is possible to use volumetric textures and textures on a transparent (with varying degrees of transparency) background. Thus, the realism of the image of such elements of the scene as vegetation, clouds, and various kinds of buildings and structures significantly increases. Realistic display of objects on the screen in Direct3D is provided with functions simulating material and light. The material in Direct3D is the surface properties of the drawn primitive. With the help of the material, it is possible to determine how light will be reflected from the surface of a 3D object. The surface properties of the object are defined in the D3DMATERIAL9 structure, where they are specified.

The package includes functions that implement the so-called morphism, i.e., the formation of periodically and smoothly changing geometric shapes and surface properties of objects in the scene. This property makes it possible to create a realistic image of the sea surface. Work with graphics in Direct3D 9 is based on matrix operations. There are three main matrices defined in Direct3D: the world matrix (World Matrix), the view matrix (View Matrix), and the projection matrix (Projection Matrix). The world matrix allows rotation, transformation and scaling of an object, and also provides each of its objects with its own local coordinate system. The view matrix sets the position and direction of viewing the scene. The projection matrix creates a projection of a three-dimensional object on a two-dimensional plane screen. With its help, the object is transformed, and the origin of coordinates is transferred to the front of the object being drawn; and the front and rear clipping planes are also determined. Sequential assignment of values for each of the matrices creates a three-dimensional scene in which it is possible to move, rotate, zoom in, delete objects of the visual scene, and perform other operations with them. When working with matrices in Direct3D, the vertex can be specified with four numbers X, Y, Z, and W in the form of a matrix, where W is not equal to 0. W is the coefficient of perspective necessary for working in three-dimensional space. In this case, the coordinates of the vertices will be equal to X/W, Y/W, and Z/W and are represented as an ordinary 4×4 matrix. The package uses the separation of data describing the elements of the external environment, and the program that provides their processing and visualization. The implementation of this principle leads to the following technology for preparing and processing data in the package: the graphic database (GDB) of the package is created using the three-dimensional graphic editor 3Dstudio MAX; GDB files are converted to x-files format available for reading by functions included in DirectX, using a converter program; when started for execution, the package loads all GDB elements into the memory and then renders (visualizes) the specified GDB fragment.

Below is a list of the main tasks solved by the package for modeling 3D images. Initialization is the registration of classes of created windows, initial installation of global variables and GDB catalogs (directories), and loading of all GDB files. Recalculation of the coordinates of moving objects, which are calculated in the ASGS with a resolution equal to the calculation cycle (usually 1 s). To simulate the smoothness of movement, it is necessary to calculate the position of objects and display them on the screen with a frame rate of at least 25 fps. Such a discrepancy leads to the need to approximate and smooth the coordinates of the points for the movement trajectory of objects. Calculation of the visual scene at a certain angle with regard to the change in the camera position—displacement or rotation around one of the three coordinate axes. Rendering—calculation, projection, and generation of the visual environment frame to be displayed on the screen. Changing the angle—displaying the displaced or rotated graphic image of the object. Hint—display of the brief help window and help—access to the package help and reference system. To recreate sound effects (noise of aircraft engines), a library of sound effects is created, from which files corresponding to the simulated aircraft types are selected in the process of take-off or landing simulation. The main MVAE hardware components are personal computers (PCs) with integrated video adapters and multimedia tools. PCs provide user interface support, geometric calculations, interaction with the ASGS over a local network, and transmission of graphic data to display devices (DD). 3D graphics functions are implemented in hardware in a specialized graphics processor–accelerator built into the video adapter, which provides hardware implementation of the DirectX 9 library functions, including support for vertex and pixel shaders of version 3.0 or higher. The selected type of display device for the MVAE may be projection systems with high-resolution projectors, plasma panels, or liquid-crystal displays. The required horizontal viewing angle is ensured by arranging the required number of screens of display devices along an arc centered at the trainee's location. When selecting a particular image tool, the following main factors are taken into account: resolution; size and shape; brightness, contrast, and chromaticity; viewing angles—horizontal and vertical; life time.

This selection depends on the functional responsibilities of the Tower controllers, the need to monitor all stages of the aircraft movement: taxiing, take-off, landing, go-around, exercising special cases, as well as the need to monitor the movement of airfield special vehicles.

9.9   Exercise Preparation Package

The software exercise preparation package (EPP) is designed to set the characteristics of all elements of the exercise and the generation of exercises for subsequent execution. An exercise is understood as a set of data on the structure of the airspace and the airfield, aircraft parameters, meteorological conditions, landing parameters and radar, graphic information elements, etc. These data constitute the ISS information base.

The main EPP functional purpose:

– in the interactive mode: introduction and correction of data using the standards of the graphical user interface used in MS Windows;
– selection of one of several exercises, work with several exercises simultaneously;
– saving exercises in the database;

– implementation of syntactic and semantic checks when entering exercise parameters and generating the internal representation;
– output of the exercise elements and messages to the operator in the diagnostics window or to a file for subsequent printing;
– generation of exercises in the internal representation and ensuring the ISS functioning in real time.

An exercise consists of elements represented in the form of various tables united in a hierarchical structure. Each specific element of the database is assigned a tabular menu, the characteristics of which are set using a submenu via dialog boxes.

The entire database is divided into a common part (general for all exercises) and an individual part for a specific area. The database has a hierarchical structure: all tables are grouped into folders by the logical similarity of information. The structure of the exercise is a set of structures of all elements of the exercise.

The information base of the EPP software consists of two main parts: universal and individual. The elements included in the universal part of the database describe the tactical, technical, and flight performances of aircraft, radar stations and various mobile objects, the text templates of most of the report tables, and various tables of symbolic names. This part of the information base does not depend on a specific area and is general (common) for various areas. The individual part consists of tables describing the characteristics of a specific air area, and exercise tables containing the structure of the simulated air situation. Area characteristics include data on aerodromes, runways, radio technical and radar locating equipment, control sectors, etc. Elements—trajectories, trajectory points, reports, and flight plans—describe the structure of the simulated air situation.

"Parameters" menu. The "Parameters" menu contains the following submenus:

1. aircraft parameters;
2. area parameters;
3. radar parameters;
4. safety intervals;
5. CDO, locals, and restricted areas;
6. mobile objects.
1. **Aircraft Parameters** The "Aircraft parameters" submenu contains the main characteristics of the aircraft necessary for the functioning of the aircraft movement model: data on fuel consumption in various flight modes, speed characteristics of the aircraft selected according to the diagrams of velocities and altitudes for various flight modes, and aircraft aerodynamic characteristics. In addition, the "Aircraft parameters" submenu contains the following aircraft characteristics necessary for air situation modeling: aircraft type, aircraft weight, wing-body geometry, roll parameters for landing, en-route and lift-off, maximum vertical speed, afterburner acceleration height, optimum afterburner pitch angle with a height greater (lower) than the afterburner acceleration height, height of the pitch angle change when climbing at top speed, optimum and maximum pitch angles above and below this height, optimum and maximum values of the angles of attack and pitch angles during descent, and characteristics of approach from the

critical point. In addition, additional data are used to simulate the aircraft movement on the ground: minimum turning radius on the ground, gear height, aircraft landing roll, aircraft run-up during takeoff in the "maximum" and "afterburner" modes, and others. Part of the data are placed in tables "Speed," "Fuel," "Aerodynamics," "Aircraft lights," and "Screw positioning".

The "Aircraft classes" element is a table of correspondence between the aircraft type and the class number. Class numbers are used when describing the approach and approach flight path fixes. The aircraft class table, in contrast to the other tables of aircraft characteristics, is located in the individual base of a specific area.

2. **Area Parameters** The "Area parameters" submenu contains the exercise parameters and the aerodrome parameters. The exercise parameters consist of the initial parameters of the control sectors and control unit call signs. The aerodrome parameters consist of runways, radio beacons, glide paths, landing systems, marker names, and data on arresting systems.

Area parameters are one of the basic elements of the exercise. This element specifies the geographical coordinates of the area center and radius. The geographic coordinates of the area center are used as the default point of reference in cases where the reference point is not explicitly specified.

The "aerodromes" element contains the names, coordinates of the airfields, magnetic declinations, transition heights, and transition levels.

Depending on the description of the aerodrome area and the adjacent regional control centers, various sectors of the controllers' responsibility (control sector) may be assigned in the exercise: Landing; CIRCLING; APPROACH; AC; LSU; TAXIING; and START.

A control sector is characterized by the following parameters: control sector name, control sector type, aerodrome number, radio channel frequency, bearing channel number, code and channel number for OLDI, radar, for which the name must be specified, and scale. For each control sector, a control sector call sign must be assigned. The call sign is used to simulate the radio communication between the controller and the pilot and includes the following: name, type of sectors (Landing, Circling, Approach, AC, LSU, Taxiing, and Start), category (civil, military, and transport), and flight type (departure, arrival, transit, and training). The initial information for the "runway" element is the aerodrome characteristics. If there are several runways in the area created, there must be available information about each of them. For each runway, the following is specified: aerodrome name, cardinal and reverse headings, coordinates of the runway center relative to the aerodrome, runway length, width, heading values, and a number of other parameters. The "Runway operation parameters" element contains signs of the runway takeoff and landing performance for each runway. Information about radio beacons and glide paths for each magnetic course is used to display the aircraft symbol on the glide path and heading indicators. Information on radio beacons is taken from the documentation for a specific aerodrome. The number of radio beacons depends on the number of runways and terrain in the aerodrome area. The "Radio beacons" element contains information on the location of the inner and outer markers, localizer and glide path beacons, and SRRNS regarding the runway. If the aircraft lands not from the final (landing) course or must perform a maneuver to take into account the individual characteristics of the aerodrome, it is necessary to set the race track direction. Information on radio beacons is set for each takeoff and landing course. The parameters of the aerodrome shall contain information about the landing systems specified in the "Landing systems" element and selected from the technical documentation individually for each aerodrome. These parameters are: system name, decision height, and minimum visual range.

For the convenience of the user, the names of the markers and the name of the runway end and center are entered in the terminology adopted at the aerodrome. The following glide path parameters shall be present in the "Glide path" element: heading code, height above the runway end, allowable deviation over the outer marker for the heading and glide path, allowable deviation over the inner marker for the heading and glide path, glide path angle, the glide path interception angle for director and automatic modes, and the distance between the lines of allowable deviations for the heading and glide path. The group of "Taxiways," "Parking places," and "Taxi routes" elements is used to build aircraft taxi routes on the aerodrome. The "Taxiways" element contains the following: way name, way type (runway (RWY), main lines, taxiway (TWY), and centerlines for vehicles), name of the airfield, information on crossings with runways, coordinates of taxiway points, signs of the report, signs of helicopter gates, marking radius, information about the taxiway width to determine the possibility of a turn in place, and the speed of the object on the taxiway. The "Parking places" (PP) element contains: the name of the parking place, the name of the airfield, type of driving in the parking place, type of the parking place, belonging to the apron, coordinates of the PP center, PP heading (direction of the aircraft nose of the at the PP), names of the taxiways with an exit from a specific PP, the radius of exit marking to these taxiways, signs of a tug call for departure or arrival at the PP, technical ways (ways for towbars), and additional parameters (they indicate the PPs through which, in the cases provided, the aircraft leaves or arrives at its PP). The "Taxi routes" element is a combination of PPs with unique names combined by the same taxi routes, used to send taxiing commands to the PP if there are several aprons at the aerodrome. It contains the following objects: route name, aerodrome name, taxi routes indicated for each runway and each takeoff and landing heading codes, consisting of taxiways leading to the runway and the main line. For the further development of the graphic image of the aerodrome, the following parameters are included in the "Area parameters": "High intensity lights", "Floodlights," and "Aerodrome marking".

3. **Radar Parameters** In the "Radar parameters" submenu, two groups of elements are defined: "Radars" of the common database and "Radars on the ground" of the database of a specific area. The common base includes tables "Control radar," "Surveillance radar," and "Landing radar." The base of the specific area includes expandable tables: "Radar setting" and "Radar closure angles." These elements contain data on the altitude and coordinates of the localizer, the surveillance period, the parameters of the visual field, the resolution of the indicator associated

with the radar in azimuth and range, and other data necessary to simulate radar information.

4. **Safety Intervals** The "Safety intervals" submenu contains elements of horizontal and vertical separation. In the horizontal separation element, the following is entered: safe longitudinal and lateral intervals for aircraft following the routes; safe longitudinal intervals when crossing flight levels (lateral intervals are the same as when moving along routes); hazardous proximity; full flight safety intervals; and forecast time. To assess flight safety, the longitudinal interval is taken into account before crossing routes, and the lateral interval—after crossing. In the vertical separation element, safe altitude differences are established.

5. **CDO, Locals, Restricted Areas** In the "CDO" (cloud data object) table, the following is set: the geometric shape of clouds (the shape that will be displayed on the localizer screen), the initial position of the center of this cloud, the shape, the distance, and the upper and lower boundaries of the cloud. If cloudiness is less than six points, then it is not displayed on the localizer. Weather avoiding is simulated in the ISS only when setting towering cumulus clouds. The "Locals" table presents local items (locals) that can hinder the movement of the aircraft. Characteristics describing locals contains their names, coordinates describing the form of a local, its maximum height. The "Restricted areas" table describes parts of the space in which aircraft operations are prohibited. The table contains their names, coordinates, and heights of the lower and upper bounds.

6. **Moving Objects** The "Moving objects" table group submenu is used to simulate various objects, such as flocks of birds, obstacles, special vehicles, etc. The group includes tables "Types of moving objects" (in the common database) and "Plans of moving objects (in specific exercises). Mobile objects "flocks of birds" are used to visually simulate the ornithological situation on the controller's station. For a more complete visual simulation of the ground situation of the aerodrome area, moving objects of types "cow," "dog," and "human" are used, as well as various vehicles (fire vehicle, ambulance car, accompanying vehicle, cleaning machine, inspection machine, truck, tractor, etc.). Each moving object has an activation time, symbolic name, type, and trajectory of movement.

"Trajectory Points" Menu The trajectory of the aircraft movement in space is described by trajectory points of various types. The "Trajectory points" menu contains the following composition of the submenus for trajectory point types: direct "Follow to" points; elevation points; turning points; coordinate points; transition points; approach points; and points of reporting.

The direct "Follow to" points, turning and coordinate points describe the projection of the aircraft trajectory on the plane. The elevation point determines the position of the aircraft in space. Approach points are used only in landing paths and set the position of the aircraft in space.

1. **Direct "Follow to" Points** The direct "Follow to" points describe a straight line segment of the aircraft movement along the trajectory. These points indicate that the aircraft flies following a previously occupied heading until reaching one of

the following parameters: the distance specified at the point; the set time at a given speed; the heading set.

2. **Elevation Points** Elevation points define a trajectory segment characterized by altitude and (or) velocity. The points are given the values of the achieved altitude, flight speed or flight speed mode (minimum, maximum, and optimal), vertical speed or path inclination angle, engine operating mode, angle of attack, and overload.

3. **Turning Points** Turning points are used to specify the turn area when the direction of the aircraft changes. At these points, the value of the heading to be achieved, the direction of the turn, and the banking angle or radius of the turn are set.

4. **Coordinate Points** Coordinate points define the coordinates the aircraft shall reach when moving from the previous point along the trajectory. A reference point (airfield) is selected, relative to which coordinates are given.

5. **Transition Points** Transition points define a segment of the trajectory by referring to another segment of this or another trajectory; they also indicate the control area taking over the aircraft movement control.

6. **Approach Points** Approach points are used only in landing paths and depend on the class and heading of the trajectory. Approach points characterize the beginning of turns in each of the landing paths (smaller box pattern and greater box pattern).

7. **Points of Reporting** These are the points where reports of a single aircraft, a group of aircraft, reports of a repeat request from a single aircraft, and a group of aircraft are made. A report is made for a single aircraft or a group of aircraft depending on whether the aircraft are grouped.

"Trajectories" Menu The structure of the exercise includes the trajectories of the routes, takeoff, arrival, and landing. Flight trajectories are built on the basis of the document describing the aerodrome area and the adjacent territory. A trajectory is a sequence of points and reports. Seven types of points provided in the EPP make it possible to describe any type of trajectory. A trajectory usually ends with a transition point. If there is no transition point, then the aircraft flies following the heading. The "Trajectories" menu contains the following submenus: route trajectories; take-off and arrival trajectories; landing paths; and track trajectories.

1. **Route Trajectories** Route trajectories are used for arriving or transit aircraft crossing the control sector. They contain the name of the trajectory with a list of points and reports. The route trajectories indicate that they belong to standard route trajectories or that the route belongs to flocks of birds. The trajectories do not depend on the airfield heading and on the classes. The route trajectories shall begin with an indication of the aircraft location in the aerodrome area. The trajectory is a sequence of segments of the path and ends with a transition point. For aircraft for which no approach is planned, the route trajectory ends with an indication of the aircraft leaving the aerodrome area and a transition point with a sign of the aircraft symbol removal.

2. **Take-Off and Arrival Trajectories** These are used for take-off and training flights, do not depend on the aircraft class, but depend on the runway heading

directions. They contain the name of the trajectory, the heading and the list of points and reports, and provide the possibility of taking off from the alternate aerodrome.

3. **Landing Paths** These are used for standard approaches. They depend on the landing heading of the runway and on the aircraft class. They contain the name of the trajectory, the heading, the class, and list of points. Landing paths can be used either for training or for arriving aircraft.

4. **Track Trajectories** These are used to set a smaller box pattern (SBP) and a greater box pattern (GBP).

"Flight Plans" Menu Flight plans are included in the mandatory part of the information database of the training exercise preparation package and contain complete information for aircraft involved in this exercise. To set a flight plan, the following information is entered in the menu table: tail number, aircraft type, flight type, flight rules, number of aircraft, equipment, wake turbulence category, route, cruising speed, arrival flight level, departure aerodrome and departure time, destination aerodrome and estimated travel time, alternate aerodromes, and other information. The listed fields in the "Flight Plan" are filled in accordance with ICAO requirements. Changing individual fields in the "Flight plans" menu enables creation of a practically new exercise.

"Reports" Menu In the process of creating a new exercise by the user, at certain defined points of the route, signs are provided indicating when reports of the simulated crew shall be given to the controller. In addition, reports shall be provided by the simulated crew at the controller's request. To implement these functions, a report database is created in the EPP. All reports are divided into groups: users; flight plans; current parameters; current landing parameters; current weather conditions; buffer; maneuvers; approach maneuvers; start and taxiing on the airfield; entry to the area; special cases; and entry of additional targets.

Reports of each group differ in the text and the list of report variables. The current information on the list of report variables is entered in the appropriate report fields during the exercise practicing. Groups of reports are divided into several subgroups. Subgroup of user reports: any changes, additions, and deletions are possible when forming reports of this subgroup. A set of user reports may reflect the specifics of flight control of a specific control area or aerodrome. The subgroup of reports on flight plans is created to ensure the issuance of reports on flight plans in the process of exercise practicing. The subgroup of reports on current parameters (current parameters, current landing parameters, and current weather conditions) enables simulation of crew reports both during flight operations along the routes and at controllers' requests. The subgroup of buffer reports enables simulation of reports from crews of arriving and transit aircraft from adjacent control sectors to simulate interaction with adjacent control sectors. Buffer reports contain information about the next point of mandatory reporting, the time of departure from it, indicate the line of entry into a given control sector and the time of passage of this line. The subgroup of message reports (maneuvers, landing maneuvers, start and taxiing on the airfield, entry to the area, special cases, and entry of additional targets) are available for changes only to the texts of the reports, but not their meaning and order. Each report is provided in a certain situation during the exercise and in a certain sequence, therefore any change in their order will result in issuing a completely different report at the right time. All reports are ordered by name, so a change of name can lead to a rearrangement of reports.

When preparing the exercises, the user opens the menu of the corresponding group and interactively adjusts the database of reports for the specific exercise.

"Weather Data" Menu The group of METEO elements characterizes the meteorological situation of this area. In the "Meteorological data" table, the parameters are set that characterize each upper-air layer, the upper boundary of the layer, wind parameters (speed and direction), weather phenomena, meteorological conditions on the runway (temperature, pressure, runway visibility, runway surface, runway friction coefficient, and humidity).

"Files" Menu The "Files" menu consists of a group of submenus that provide the user with additional functions: selecting a specific exercise from the list; performing the procedure of generating internal presentation files for later replaying in the ISS; exercises printing; test area printing; planning tables printing; color settings; data validation; and exiting the package.

"Help" Menu The "Help" menu provides on-line help to the EPP user in the process of preparing new exercises, and the user can be provided with full information on working with the EPP using the dialog box-based interface.

9.10 Planning Information Simulation Package

The planning information simulation package (PISP) is designed to ensure the preparation, adjustment, processing, and storage of planning information in the ISS. The PISP organization is based on the preparation of a set of flight plans with the subsequent formation of a planning table, on the basis of which the aircraft simulation is activated in real time. Flight plans are prepared in the preparatory phase of the ISS using the methods of the exercise preparation package (EPP). Each flight plan contains the following basic service information: identification index; transponder code; rules and type of flight; aircraft type; aircraft equipment; airfield and time of departure; cruising speed and flight level; route; destination aerodrome and total estimated time; and alternate airfields.

The prepared flight plans are transmitted to the RPIPS for organizing the simulation of daily planning with the subsequent issuance of planning information to the AWS-C. In the process of training, the planning information can be corrected in terms of changes in the time of entry (takeoff), the flight number, the transponder code, and the addition or cancelation of a flight. The corrected information is also processed and generated at the RPIPS. Flight plans are included in the mandatory part of the information base of the exercise preparation package software. The flight plan contains comprehensive information for a specific flight. The necessary information for specifying the flight plan is the following: time, flight (tail) number (five or seven characters), aircraft type, aerodrome number, take-off and landing runway number, flight type, ground trajectory, take-off trajectory, arrival trajectory, landing trajectory, depending on the flight type, fuel availability, and cleared flight level. In Fig. 9.4, a flight plan table is presented in the form of a dialog panel implemented using the dialog box technology with drop-down menus. As shown in Fig. 9.4, there is a set of "Select" keys that facilitate filling in the flight plan fields. For example, when the route "Select" key is pressed, it highlights all the trajectories of the routes created in this exercise. Flight plans are created, as a rule, at the final stage of the information base generation, since the main part of the base elements is used to create them, and it is convenient to use the entire service offered by the training package to enter them.

The tabular form of the flight plan (Fig. 9.4) shows the fields of the ICAO FLIGHT PLAN.

At the same time, without changing other elements, and by changing only individual flight plans or some fields in them, a practically new exercise can be obtained.

These fields are filled in accordance with ICAO requirements. To fill in the "Route" field, the "Insert" key of the element menu can be used: take-off trajectories, route trajectories, arrival trajectories, and coordinate points; it is also possible to enter the route parameters from the keyboard. To fill this field, the trajectories of the routes can be used in full or partially. In this case, it is necessary to enter points of entry into the trajectory segment and points of exit. If the point is the intersection point of two trajectories, it is entered once. The parameters of the route field enable changing the speed and cruising level of the aircraft when it reaches a specific point in the aircraft position.

Cruising level and speed are set by ICAO rules. The aircraft position is set by the coordinate point from the menu, or by setting the azimuth and range from the keyboard.

For departing aircraft, the route begins with the name of the take-off trajectory per ICAO (regardless of the take-off and landing heading) corresponding to the name of the specific take-off trajectory. For arriving aircraft, the route field ends with the name of the standard ICAO arrival trajectory (regardless of the heading) corresponding to the name of the specific arrival trajectory. When practicing the exercise in real time, the current heading may be changed.

For transit and arriving aircraft, the route description begins with a route trajectory with entry and exit points. When a route name is selected from the menu or is typed in the window, the name itself is selected without underlining. Trajectory names with underlining are used to indicate that the trajectory has a paired trajectory in the opposite direction. During the exercise, the air situation model automatically determines the direction of the trajectory from the entry and exit points. The data on the planning information are stored in the file of the external presentation of the database with the name "Planning table" presented in Fig. 9.5.

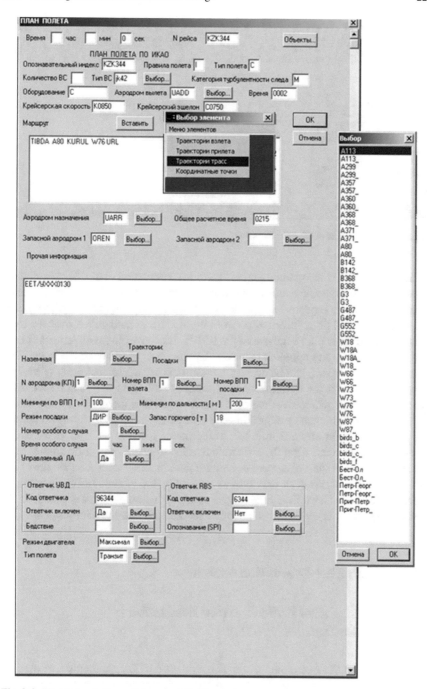

Fig. 9.4   Flight plan table as dialog panel (in Russian)

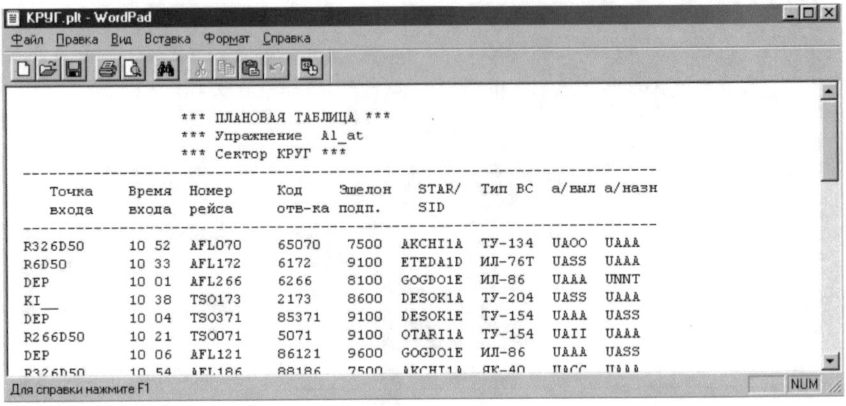

*** ПЛАНОВАЯ ТАБЛИЦА ***
*** Упражнение A1_at
*** Сектор КРУГ ***

Точка входа	Время входа	Номер рейса	Код отв-ка	Эшелон подп.	STAR/ SID	Тип ВС	а/выл	а/назн
R326D50	10 52	AFL070	65070	7500	AKCHI1A	ТУ-134	UAOO	UAAA
R6D50	10 33	AFL172	6172	9100	ETEDA1D	ИЛ-76Т	UASS	UAAA
DEP	10 01	AFL266	6266	8100	GOGDO1E	ИЛ-86	UAAA	UNNT
KI__	10 38	TSO173	2173	8600	DESOK1A	ТУ-204	UASS	UAAA
DEP	10 04	TSO371	85371	9100	DESOK1E	ТУ-154	UAAA	UASS
R266D50	10 21	TSO071	5071	9100	OTARI1A	ТУ-154	UAII	UAAA
DEP	10 06	AFL121	86121	9600	GOGDO1E	ИЛ-86	UAAA	UASS
R326D50	10 54	AFL186	88186	7500	AKCHI1A	ЯК-40	UASS	UAAA

Fig. 9.5 Planning table display (in Russian)

The option "Print the planning table" provides preparation of a table containing a list of aircraft in a specific control area, an entry point and time of entry into this area, arrival or departure trajectories, and arrival–departure aerodromes. Before running the exercise in real time, information from flight plans is transmitted over the local network to the flight plan database of the RPIPS. It is possible to implement at least 200 flight plans in one complex exercise. The duration of the exercise in the RTE (real-time environment) can be arbitrary. From the methodological point of view, it is advisable to limit the duration of the exercise to 180 min with a corresponding set of flight plans, where the time of the exercise start is indicated. Routes (route trajectories) in flight plans are built in such a way that symbols of simulated aircraft are displayed no less than 5 min before the passage of the point of entry to the simulated control area. After the aircraft leaves the simulated air traffic control area, the aircraft symbol shall disappear not earlier than in 5 min after the passage of the exit point, unless it has been previously removed by the operator pilot (instructor). After a certain predetermined time (e.g., 5 min), upon completing the practicing of the last flight plan, this exercise is re-run in accordance with the planning table.

9.11 Automated Evaluation System

9.11.1 Information Collection and Evaluation in the Training Process

The structure of the automated evaluation system (AES) organization in modern simulators for air traffic controllers is based on the development of the current base of monitored parameters [1, 2] in the course of the training with subsequent comparing of them with standard parameters from the prepared base of standard parameters. The structure of the AES is presented in Fig. 9.6.

The current database of monitored parameters is implemented on an open basis of a constantly evolving structure. Considering the versatility and complexity of the functional tasks solved by air traffic controllers, in the ranking of skills, convolutions of the composition of monitored parameters are used taking into account various restrictions. The main stages in professional skills ranking are their specialization by control sectors and functional tasks methodologically divided into groups of activity parameters: interaction with adjacent control centers; graphic control; aircraft identification and positioning; observance of the rules of radio communication and radio phraseology; compliance with the ATC transmit-to-receive lines; decision making on aircraft movement control; compliance with established routes of aircraft; observance of separation intervals; hazardous proximity of aircraft (near midair collision); controller's actions under special conditions; and special flight situations.

For each group of activity parameters, a set of monitored parameters is created, each of which shall be characterized by physical, quantitative values, or represented by a certain algorithm of its description. The monitored parameters used in the

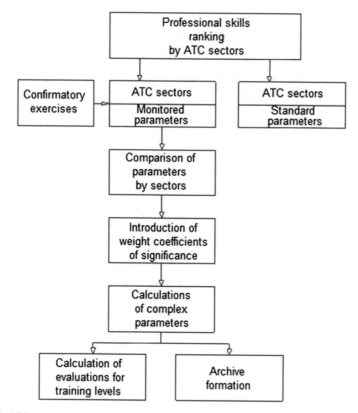

Fig. 9.6 AES organization structure

simulators for air traffic controllers shall meet a number of requirements, the main of which are:

- completeness—all basic skills shall be controlled, i.e., a formalizable relationship shall be established between the main skills from their list and the parameters to be monitored;
- reliability—they shall adequately reflect the corresponding parameter using the obtained evaluations;
- accuracy—the monitored parameters shall be measurable and evaluated with reasonable accuracy;
- purposefulness—the requirements of the adequacy of the skills being practiced shall be observed;
- scalability—standard values of monitored parameters shall be defined in such a way that they correspond with the real activity of the controller, i.e., that the requirement of adequacy of the applied standards to real indicators of the controller's work is observed;
- feasibility—measurements and calculations of monitored parameters shall not cause fundamental difficulties when using standard tools of modern training equipment;
- simplicity—the number of monitored parameters shall be minimal and optimal provided that the previous requirements are met.

When creating and developing the AES, it is necessary to organize and improve the database of standard parameters differentiated by control sectors and by groups of skills practiced during the exercise on ATC simulators. Standard parameters are defined and substantiated: by the use of guidance documents for the preparation of air traffic controllers; by statistical processing of data obtained during the repeated execution of confirmatory exercises by various groups of trainees; by the method of expert evaluation. Comparison of the current monitored parameters with the relevant standard parameters for a given ATC sector makes it possible to obtain a set of particular indicators characterizing the controller's level of training according to various criteria.

To obtain the total values for the complex criterion of the level of training, it is necessary to apply weight coefficients to each particular indicator that characterize the measure of the significance of each indicator.

The basic values of the weight coefficients of significance for particular indicators are determined by expertise and can be changed by the administrator of the simulator depending on the objectives of training at various stages of training.

Thus, the calculation of evaluations for the controller's level of training during the confirming exercise k can be represented as follows (9.8):

$$U_k = \sum_{n=1}^{S} W_n^k I_n, \qquad (9.8)$$

where W_n^k are weight coefficients characterizing the measure of the significance of the nth particular indicator during the kth exercise, $k = \overline{1, m}$; I_n are particular indicators that determine the controller's level of training.

Here, the following conditions are always observed:

$$\sum_{n=1}^{S} W_n^k = 1, \quad W_n^k \geq 0. \tag{9.9}$$

The resulting level of training will be determined by the sum of evaluations for the training course, consisting of m confirming exercises:

$$U_o = \sum_{k=1}^{k=m} U_k. \tag{9.10}$$

For clarity, the controllers' level of training can be presented using the normalizing of particular indicators relative to some limit values of particular indicators I_n^p:

$$I_n^{\text{\tiny H}} = I_n / I_n^p, \quad \text{with} \quad I_n^p \geq I_n \quad \text{and} \quad 0 \leq I_n^p \leq 1, \tag{9.11}$$

where $I_n^{\text{\tiny H}}$ is the normalized value of the particular indicator.

The number of particular indicators used in determining the level of training of the control staff can be supplemented and improved in the process of methodological support during the operation of training packages.

9.11.2 Aircraft Conflicts

Training systems are a unique tool for practicing skills to resolve conflict situations. In the preparatory phase of the ISS functioning, in accordance with the methodological decisions of the training supervisor and based on the use of the exercise preparation package, training exercises can be created, in which conflict situations occur during the process of training. At the same time, one exercise can include an arbitrary number of conflicts of various kinds:

- hazardous proximity of aircraft (near midair collisions);
- violation of the separation standards;
- violation of the lateral separation intervals when following the same route;
- violation of the vertical separation interval between aircraft flying on the same flight level;
- violation of the longitudinal separation interval and the lateral separation interval when overtaking;
- violation of rules and standards when crossing an opposite traffic flight level;

– violation of rules and standards when crossing an in-trail traffic flight level;
– violation of rules and standards when aircraft fly along intersecting routes;
– violation of rules and regulations of aircraft movement at the aerodrome.

The presence for conflict situations is checked at each cycle of the calculation in the air and ground situation model, after calculating the current coordinates of the aircraft $\{x_C; y_C; H_C\}$, the heading vector $\vec{\varphi}$ and the velocity vector $\vec{V}$. Comparing, for example, the current coordinates of the aircraft with the coordinates of the nearest points lying on the planes of the boundaries of the vertical (top, bottom) $\{x_C^T; y_C^T; H_C^T\}$, $\{x_C^B; y_C^B; H_C^B\}$ and horizontal (right, left) $\{x_C^R; y_C^R; H_C^R\}$, $\{x_C^L; y_C^L; H_C^L\}$ separation makes it possible to obtain the values of deviations from the separation boundaries R_C. When $R_C^T \geq 0$ or $R_C^B \leq 0$, violation of height separation is registered. When $R_C^R \geq 0$ or $R_C^L \leq 0$, violation of horizontal separation is registered. To detect individual conflicts, standard parameters shall be used that are stored in the database. For example, when registering a conflict when overtaking, it is necessary to take into account such a standard parameter as compliance with the lateral interval (not less than 10 km), if the minimum standard longitudinal interval is not maintained. A detailed analysis of one or a group of conflicting aircraft makes it possible to display the following data: conflict time, cause of the conflict, tail numbers of the conflicting aircraft, azimuth, range, coordinates of the conflicting aircraft, flight heights, headings, ground and vertical speeds, distance between the aircraft, time-to-collision and closing speed. It is possible to print this information and to display a graphic image of conflicting aircraft.

9.11.3 Processing and Documenting the Training Results

The automated evaluation system (AES) provides for the collection of specific information in the course of the exercise, identifies, and records the conflict situations that have arisen due to the controller's fault, as well as the processing and evaluation of the trained controller's results. In the course of the training, all the necessary data are collected and recorded on the hard disk. The record of each exercise is identified by the name of the current scenario, the date, and time of the exercise. At the time of starting, the user also can enter any symbolic name to further identify the current record. The recorded information is transmitted from the ISS workstations and stored in a database on group equipment. When starting the AES, the user can select the record of the results of the desired exercise. For each control sector involved in the selected record, the user can obtain information about the monitored parameters. To analyze the results of exercises, the user is provided with a call for detailed information on incorrect actions, for example, violations of the rules and standards of flight operations: violations of safe intervals between aircraft, flight through danger and restricted areas, violations of safe flight heights, etc. For each case, the following is displayed: time of violation, type of violation, flight (tail) number, and current aircraft parameters. For a group of conflicting aircraft, some parameters are also displayed: time of the conflict, cause of the conflict, tail numbers of the conflicting aircraft, coordinates of the conflicting aircraft, flight height, heading, ground and vertical speed, distance between the aircraft, time-to-collision, and closing speed. It is possible to print the necessary information on the printer. For a detailed analysis of conflict situations, it is possible to examine video fragments in a window simulating the air situation display that shows the aircraft flight paths. For visual presentation of information in methodological support of the simulators, uniform forms of control and reporting documentation (tables, protocols, and reports) are developed, which contain information about the results of the exercise: number of aircraft serviced; time the aircraft spent in the sector; data on aircraft that have run out of fuel during the flight; data about the aircraft performing a missed approach without the relevant command; list of aircraft that flew beyond the radar visual field; information about the commands entered by the instructor for this aircraft; texts of reports of the selected aircraft; unanswered aircraft requests; and time interval between the request and the response in the case when it exceeds the acceptable value.

The AES package includes functions of accumulation and archiving of training results with the possibility of printing the results for individual trainees and various groups of trainees. The tasks of archiving also include systematization of the training (control) results by various criteria (by age, by work experience, by the number of approvals at control centers, etc.) in order to further improve the methodology for training the control personnel.

9.12   Future Prospects

The human factor is the cause of 70–80% of all accidents in civil and military aviation. The most important player, which largely determines the reliability of the air traffic control system functioning as a complex human–machine system, is the air traffic control system controller [2]. The general trend of increasing the reliability of the air traffic controller as the most important element of the ATC system is a wider introduction and use of practical simulator training to improve their professional level. In this regard, the development and improvement of air traffic controller simulators is a very topical issue, which largely determines the flight safety [3, 4].

Analysis of the developments of leading companies in the development and implementation of air traffic controller simulators reveals the following main trends and directions in the development of air traffic controller training systems:

1. Implementation of the modular flexible architecture of the simulator according to the principle of an open information system ensuring the invariance of the configuration of functional modules from combinations of the same type to an arbitrary combination depending on the structure of the simulated ATC area and the composition of the control sectors. Information models of the training system subsystem (AWS-C) shall correspond to a specific type of a simulated ATC system. The modeling subsystem is built according to the universal principle and

provides for the modeling and generation of information on the air situation from all types of radar stations, ADS, ARs, OLDI channels, CPDLC data transmission links, and planning and meteorological information, regardless of information sources.

2. The increasing level of performance of technical means of information technology allows increasing the adequacy of modeling air and ground situations, especially under unusual situations: in special cases, special conditions and emergency situations.

3. Further development and implementation of operational objective monitoring for the actions of the trained controllers and automated planning of training based on an individual professional evaluation of the trainees with the definition of requirements for training exercises (and their automated initialization). Introduction of automated control system software is to assess the quality level of the controllers' handling competence by the training results.

4. Improving the efficiency of controller simulators by automating the functions of operating personnel (gradual exclusion of operator-pilots from the control circuit of the ATC system model) through the introduction of modern technologies in the field of speech synthesis and recognition.

5. Development and improvement of the adequacy of modeling the visual aerodrome environment with control of aircraft on the airfield through the introduction of the latest applications in the field of creating 3D images using 3D Studio Max editor with dynamic libraries OPEN GL, DirectX API based on the latest computer technologies.

6. The accumulated experience allows synthesizing air traffic controllers simulator as expert systems in the future. Figure 9.7 shows the structure of the air traffic controllers simulator implemented on the principle of an expert adaptive system with the means ensuring understanding the natural professional language of the air traffic controller and speech synthesis, which has a database and knowledge base, with an automatic evaluation system for the controllers' level of professional training and an adaptive planning system to adjust the complexity of exercises.

Consistent implementation of these areas in the field of creation and implementation of training systems ensures more efficient training of air traffic controllers and is one of the most important ways to improve flight safety.

Section Review

1. The structure and configuration of the special software of the integrated system simulator for air traffic controllers.

2. Methods for modeling the aircraft movement, automatic generation of the movement trajectories for moving objects in the integrated system simulator for air traffic controllers.

3. Modeling radio-technical means of flight operations support in the integrated system simulator for air traffic controllers.

4. Simulation of RTF FS faults in simulators for air traffic controllers.

5. Principles of meteorological modeling in simulators for air traffic controllers.

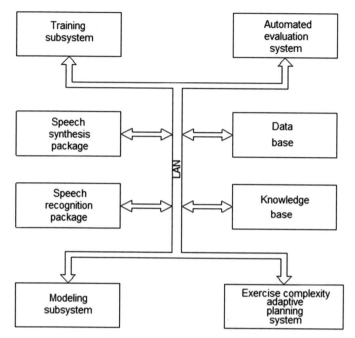

Fig. 9.7   ISS structure as an expert-adaptive system

6. Simulation of the visual aerodrome environment in simulators for air traffic controllers.
7. Organization of the exercise preparation package database.
8. Organization of planning information modeling in simulators for air traffic controllers.
9. The automated evaluation system to assess the level of handling competence in simulators for air traffic controllers.

References

1. Filin AD (1983) Method for estimating errors in simulating the flight path of targets in digital ATC simulators. Air Transport, Moscow
2. Kovalenko GV, Kryzhanovsky GA, Sukhikh NN, Khoroshavtsev YuE et al (1996) Automation of air traffic control processes. Air Transport, Moscow, p 320
3. Filin AD et al (1991) Promising simulators for air traffic controllers with the automation of operational service personnel's functions. Leningrad: Radio Electronics Issues, Issue 19, pp 56–61
4. Filin AD, Suleimanov RN et al (2010) Studying ways to create technical means to ensure flight safety, to assess the level of combat training in Air Forces subunits and units during trainings, exercises and live firing. Research report (stage III). VNIIRA, SPb, p 271/84

Chapter 10
ATC Radiotechnical Aids

10.1 ATC Radars

In order to perform the ATC procedures, the controller requires data on actual aircraft location within the area of responsibility. The main source of such data is radars. At different movement stages, different radars are used: when at the aerodrome surface—the airfield surveillance radars; when flying within the aerodrome area—aerodrome surveillance radars; and when on route—route surveillance radars. The ATC radars use two coordinates: They determine the aircraft azimuth and slant distance so that after the radar data processing the controller receives horizontal coordinates of the aircraft [1]. For ATC tasks, both primary and secondary radars are used. Primary radars measure the coordinates based on the signals reflected by the aircraft fuselage. They are fully self-contained as they do not require any onboard or other additional equipment. Secondary radars (radars with active response) require cooperation with the onboard transponders, but they provide more data on the object under surveillance. They are used both independently and jointly with secondary radars; in this case, the combination of the primary and secondary radars is called integrated system complex (ISC). The ISC can also include the primary processing equipment, radio communications equipment, automated direction finder (usually, as standby means). Antenna system, primary processing equipment and other equipment can be the joint for the primary and secondary ISC channels [2].

Depending on the ATC radar coverage, they are classified as on route and aerodrome.

The airport surveillance radar (ASR) is used to get the air situation data in order to provide the ATC within the aerodrome area and to bring the aircraft into the landing aids coverage. The ASR coverage by distance depending on the class is 150, 80, and 46 km. The azimuth is used to provide the round-looking scan, and the elevation is used to provide the scanning with the angle range from 0.3° to 45°. In order to provide such coverage area, an antenna with directional pattern rotating in horizontal plane is used; this directional pattern is narrow in azimuth (about 1.5°) and wide in elevation (in elevation plane, the directional pattern has the form of a

© Springer Nature Singapore Pte Ltd. 2020
Bestugin A.R. et al., *Air Traffic Control Automated Systems*,
Springer Aerospace Technology, https://doi.org/10.1007/978-981-13-9386-0_10

cosecant). The scanning rate is 10–15 rpm. The resolution by azimuth is 1.5°–2°, and by distance—400 m. The precision of the coordinates' measurement by azimuth is 12′, and by distance—300 m. The ASRs work in within the range of 3, 7, 10, 12, 20 cm [3]. In order to fight the passive interference (reflections from fixed objects), the moving target indication (MTI) mode is used, which is based on the frequency filtering of the incoming signals (signals reflected from the aircraft are shifted by frequency due to the multiple reflections from the fixed objects caused by the Doppler effect). In order to reduce the influence that hydrometeors have on the radar operation, a change of the type of the emitted signal polarization is used (from linear to elliptic).

The enroute surveillance radars are designed to provide the radar control of the airspace outside the aerodrome area and on the routes, as well as to get the weather data. The enroute surveillance radars are subdivided into two classes: with the range of about 400 and 250 km. The minimum aircraft detection distance is from 5 to 15 km, and the altitude—20 to 30 km. The coverage area by azimuth is round with the scanning rate not less than 5 rpm, and by elevation—from 0.3°–0.5° to 35°–45°. The resolution of the radar by distance is 1 km, and by azimuth—about 1.5°. The precision of the distance measurement is about 0.5 km, and by azimuth—about 0.5° [4].

Secondary radars are using the active answer principle: The ground ATC radar is emitting the interrogation signal; the onboard equipment is receiving it and is re-emitting the feedback signal in the required format. Thus, the required coverage area can be ensured with the equipment of less power potential. Besides this, in the feedback signals the aircraft are passing the identification data and other information required for the air traffic management. By range, the secondary radars can also be subdivided into aerodrome (100–200 km) and on route (up to 400 km). The minimum range is not more than 2 km, and the scanning within the elevation plane is from 0.5° to 45°. The interrogation signals are emitted through a narrow rotating directional pattern (about 3°–4°) by azimuth and wide in the elevation plane. Antenna system of the onboard transponders has the directional pattern, which is omnidirectional in horizontal plane and low directive in the vertical plane. Within the secondary radar systems, the data transfer is realized by pulse coding. In request codes (uplink) the interval-time dual-pulse coding, which means that each value is corresponding with the specific coding interval. The response codes (downlink) are formed by position coding method: Each symbol is transmitted by the absence or the presence of a pulse at a specific point of time [2].

The secondary radar systems can function in "ATC" mode (national systems) and "RBS" mode (the systems answering the international standards). They differ in the operating frequencies of interrogation (837 MHz in "ATC" mode and 1030 MHz in "RBS" mode) and response (740 MHz in "ATC" mode and 1090 MHz in "RBS" mode) signals, as well as in the message coding format. The interrogation signals are mostly the identification interrogation (coding range of 8 μs for "RBS" "A" mode, 17 μs for the "RBS" "B" mode, and 9.4 μs for the "ATC" mode), and interrogations on the aircraft movement parameters (coding range of 21 μs for "RBS" "C" mode and 14 μs for the "ATC" mode) [5]. Within the "RBS" mode in the response codes, the altitude is passed, and in the "ATC" mode—the altitude, the remaining fuel, possibly the heading, the aircraft speed, and the event signals. In order to serve the aircraft with transponders in either mode with only one radar, interrogation codes of a combined mode are generated.

Besides this, in order to free the frequency band, which is designed or international standards of communication equipment (which includes the operating frequencies of the secondary radar in "ATC" mode), the measurements for full or partial translation of the interrogation and feedback frequencies of secondary radars, which are operating based on national standards.

In order to avoid mistakes in azimuth determination (due to the signal passing the minor lobes), specific preventive measures are introduced. Besides the main dual-pulse interrogation coding, emitted through the main antenna, in the pause between the pulses a third pulse is emitted through the compensation antenna. The directional pattern of the compensation antenna is low directive; it has a minimum toward the maximum of the directional pattern of the main antenna and exceeds its level in the area of the minor lobes of the directional pattern of the main antenna.

Thus, when comparing the amplitude pulses received by the onboard equipment, the decision is taken by which lobe the coded transmission is passed. If the transmission was done through the minor lobes, the transponder is generating the feedback transmission. This is the principle of the on-demand suppression of the minor lobes (see Fig. 10.1). When receiving the feedback signal, the secondary radar also implements the principle of the on-demand feedback suppression of the minor lobes: When analyzing the ratio of the signals in the main and compensation directional patterns, the decision is taken on either to count the received feedback or to ignore it (taking it as a false one passing the minor lobe).

The feedback code "RBS" consists of two framing pulses separated by 20.3 μs and 12 positions, evenly distributed at every 1.45 μs. The pulses are emitted at the required positions, thus passing the four-digit octal number, which characterizes the registration number or altitude depending on the received interrogation. Besides this, following the uplink radar interrogation (and by pressing a button) a special identification pulse is emitted after 4.35 μs after the second framing pulse.

The registration numbers are assigned by the controller's command when the aircraft enters his/her area of responsibility. The aircraft, which do not have an assigned code, the 20008 number code is assigned. Besides this, by means of transferring some singled out combinations of four-digit octal numbers, the service signals are forwarded, for example, "loss of contact," "emergency," and "unauthorized aircraft use". Within the "RBS" mode, the altitude is coded in increments of 100 feet up to the value of 126,700 feet.

The feedback "ATC" codes contain the following parts: coordinate, key, and information. The coordinate code is dual pulse and is designed to get the symbol at the radar display. The key code is triple pulse; it defines the feature of the data that is passed forward: registration number (key 110) or current information (key 000)—altitude, residual fuel, and event signals. The key and the data are transferred by the position method with the active pause. The active pause method entails that both "1" and "0" are transferred with a pulse, and for "1" and "0" different time positions are assigned

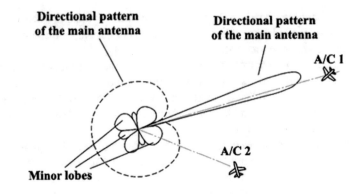

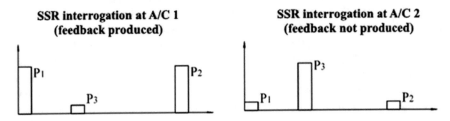

Fig. 10.1 Suppression of the minor lobes of the secondary radar

(each digit corresponds to two time positions). Such a method gives better results in terms of the interference resistance.

The registration number is transferred as five dual quadruples, which is a five-digit decimal number (aircraft registration number is assigned once when the aircraft enters into service). The current information is transferred as follows. The altitude is coded by three full quadruples and two digits of the fourth quadruple (the altitude value is transferred to a precision of 10 m within a range from 500 m to 30,000 m).

The fourth digit of the fourth quadruple of the binary units is assigned to the pressure feature, which shows "1" in case if the reference pressure on the pressure altimeter is set to 760 mm Hg (by default), and "0" when the reference pressure is set same as the pressure at the destination airport.

The fifth digit of the fourth quadruple of the binary units is assigned to the "emergency" feature. The fifth quadruple of the binary units characterizes the residual fuel in the fuel tanks as percent of their full capacity. In Figs. 10.2 and 10.3, the structures of the interrogation and feedback codes of the secondary radar in different modes are shown.

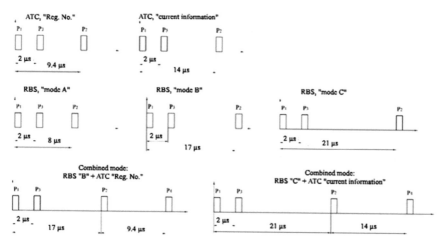

Fig. 10.2 Schematic view of the secondary radar interrogation codes

10.2 ADS System

The conventional air traffic control system assumes that the main source of the navigation data is ground radars, at that the aircraft position can be known to the controller to a more precision than to the flight crew of this aircraft. Such disposition changes with the development of the second generation of the satellite navigation. When the global satellite system aids are used on board of the aircraft, it gives the opportunity to define the coordinates to a tolerance of about 30 m and frequency of about 1 Hz, which means that the precision of the positioning is significantly higher than the precision of the position determination with the help of the ground radio navigation and radar aids. Besides this, when the ground and satellite additions to the global system are functioning, in the differential mode of the satellite navigation the precision of the positioning goes up to ones and even fractions of meter. Also, the modern aircraft provide for complexing of information, which is received from different probes, within the integrated flight and navigation system [6]. Within the integrated flight and navigation system the information from the ground radio navigation systems, self-contained aids and satellite navigation system (a rough structure of the integrated flight and navigation system is shown in Fig. 10.4).

The air data computer (ADC) system calculates the aircraft speed vector; the inertial navigation system (INS) calculates the aircraft coordinates with the help of independent aids (by the data received from gyros and accelerometers); the pressure altimeter calculates the altitude referenced to a preset constant pressure surface; the Doppler system calculates the speed and the drift angle by radar methods (based on the shifting of frequencies of the signals reflected from the ground surface); the low-altitude and the high-altitude radio altimeters measure the heights to the relief of the underlying surface by radar means; the marker receiver (MKR) defines the time of

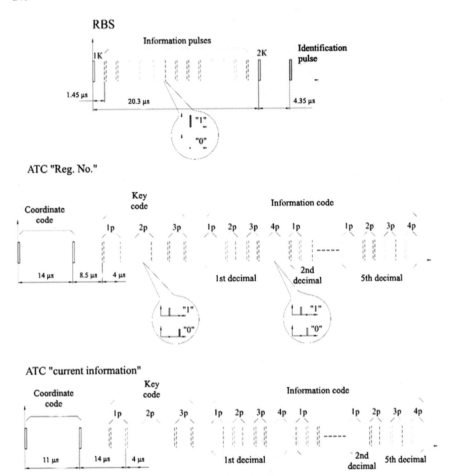

Fig. 10.3 Structure of the feedback signals of the secondary radar

passing of the marker beacon; the aircraft transponder is part of the secondary radar system; the VHF transceiver serves for acquisition of the differential corrections of the satellite navigation system and data exchange within the ADS system; automatic direction finder (ADF) shows the relative bearing of communication radio and locator beacons; the short-range radio navigation system equipment defines the azimuth by the navigation field of ground beacons; the distance measuring equipment (DME) performs the function of an interrogator of the ground radio relay station; the glide slope receiver (GS REC) and the localizer receiver (LOC) define deviations from the glide slope in two planes using the instrument landing system (ILS); the receiver of the satellite navigation system (SNS) receives the signals from the navigation satellites and calculates the geographic coordinates of the aircraft; the flight digital

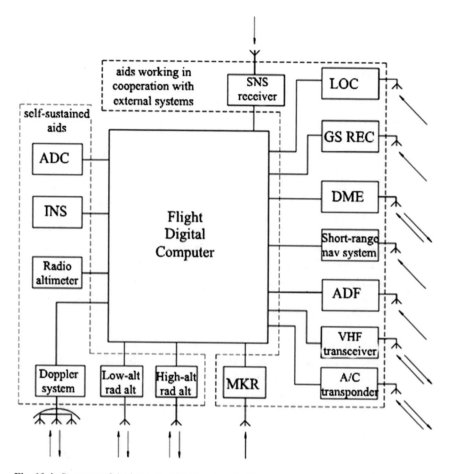

Fig. 10.4 Structure of the integrated flight and navigation system

computer performs complexing of all received data and processes it in accordance with the preset algorithms.

Following such approach allows filtering of navigation measurements and avoiding abnormal (serious) errors. The reliability of the received data significantly increases due to the uncorrelatedness of the errors of the measurers of a different nature. Besides this, between the measurements it is possible to use corrected indications of the inertial system (by the satellite signals), which means that finally the integrated flight and navigation system output sufficiently precise coordinates almost continuously. Thus, quite consistently a trend appeared to transfer the precise coordinates, which are calculated on board, to the ATC controllers. In order to implement such broadcasting, a digital radio communication system is required. The concept of the automated downlink and uplink radio communications (taking into account that not only the coordinates, but all other required information can be transferred) was considered to be an advanced ATC provision aid. Such an approach was taken as a basis for the automatic dependent surveillance (ADS) system [7]. Currently, there are two types of ADS systems: contract and broadcast.

The ADS-contract (ADS-C) system offers the data exchange only between the traffic participants that have signed an agreement on data receiving. The message parameters are then defined by the agreement. The ADS-C system is in a certain way similar to the secondary radar, but its operation is performed not to the omnidirectional antenna, and the interrogations are addressed in order to allow for differentiating the feedback from different aircraft. By the uplink, the formalized messages, which substitute the controller's voice commands, can be transferred.

The ADS-broadcast (ADS-B) system offers the addressless transfer of data from the aircraft, which means that the data are broadcasted to all concerned traffic participants via the VHF data link. All transceivers are tuned to one frequency that is why the principle "everyone sees everyone" is realized. The differentiation of message from multiple aircraft is done by the principle of time. The possibility to ensure the channel differentiation by time is simplified by the presence of universal time reference within each ADS transceiver (because of the use of satellite receiving measurers). The whole time resource is divided into quanta. By analyzing the broadcast content, each new communications participant defines those quanta, which he can occupy without creating any conflicts. Thus, the principle of time differentiation is similar to the principles used in mobile communications, but at the same time this channel is self-organized due to the presence of the precise time reference at each aircraft, which means that it does not require general management or synchronization. In this respect, the ADS base station is functionally similar to the ADS transceivers located on board. The ADS system operation principle is shown in Fig. 10.5.

Besides the control commands, weather data, and satellite navigation system differential corrections, via downlink channel the data on the coordinates of aircraft are not equipped with the ADS transceivers and are detected by the ground radars. Thus, on board of each aircraft the air situation data almost identical to the controller's data can be received.

10.3   The Short-Range Radio Navigation Systems

The short-range radio navigation aids are designed to guide the aircraft to the aerodrome area and to define the horizontal coordinates of the aircraft on route. The aircraft coordinate measurement is conducted by joint operation of the onboard equipment and ground beacons, which have a certain known-to-everyone position (each ground beacon is assigned to its own operating frequency and the identification code). The onboard equipment measures the azimuth and slant distance to the ground short-range navigation system position, and by the results of such measurements the integrated flight and navigation system can calculate the aircraft geographic coordinates. The coverage of a standard short-range navigation system is 400 km.

Fig. 10.5 ADS system
functioning principle

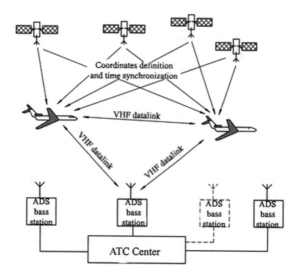

ADS system structure

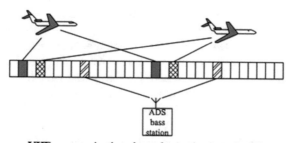

VHF communication channel organization principle

10.3.1 VOR/DME System

The international short-range radar navigation is provided by a combination of two systems: VOR and DME. The very-high-frequency omnidirectional radio beacon (VOR) system is utilizing the azimuth channel, and the distance measuring equipment (DME) system provides the distance measurement.

The very-high-frequency omnidirectional radio beacons operate in 108–118 MHz frequencies (with intervals of 100 kHz between the frequencies) with horizontally polarized emission. The VOR operation principle consists in the following: The ground beacon emits two radio signals: a reference and an azimuth signal. The reference signal is emitted through the omnidirectional antenna (with a round directional pattern in horizontal plane).

The reference signal (or a "constant phase" signal) is generated due to an amplitude modulation of a high-frequency oscillation by a signal with the frequency of 9960 Hz, which in its turn is modulated by a frequency of 30 Hz. Besides this, the reference signal provides the identification function: The data on the number and coordinates of the beacon are generated by a manipulation with an audio frequency of 1020 Hz (the Morse code is used).

The azimuth signal (a "variable phase" signal) is emitted through a directive antenna with a directional pattern in a horizontal plane in the form of a cardioid, which is turning clockwise at a rate of 30 rps.

The antenna of the rotating field consists of two crossed dipoles, which are fed by signals with a phase difference in a frequency of 30 Hz equal to 90°.

Thus, at the receiving point the signal emitted by a directive antenna looks like an amplitude-modulated oscillation with the modulation frequency of 30 Hz, whose phase depends on the azimuth, where the receiver is located.

By comparing the phases of two first channels in a phase detector, the azimuth value is defined, and the third channel gives information on the beacon identification parameters. The azimuth measurement precision is about 4°–5°. The VOR system operating principle is shown in Fig. 10.6.

The reference and azimuth signals are received by the onboard equipment and are subjected to an amplitude detection; after this, a separate processing in three filters with the frequencies of 30, 9960 Hz (within the channel with a bandwidth frequency of 9960 Hz, the amplitude restriction and frequency detection are also provided in order to differentiate the "reference phase" signal), and 1020 Hz is provided.

In order to make the azimuth determination more precise, the VOR Doppler system is used. As the rotating field antenna, a ring-phased antenna array is used with the radius of about 10 m with the directional pattern rotating rate of 30 rps. Through it, the azimuth signal (a "variable phase" signal) is transmitted in the form of an amplitude-modulated carrier with a frequency of 9960 Hz, which is modulated by frequency with a 30 Hz oscillation with a deviation ratio of 16.

Which means, that from the user's point of view the reference and azimuth signals of a regular and Doppler VOR are switched. Due to the fact that the Doppler VOR adds the data into the azimuth signal with the help of a frequency and not an amplitude modulation, the system is more protected from distortion due to the reflection from the local objects. When performing an amplitude, and then also the frequency detection of the received azimuth signal within the onboard equipment, by its phase relative to the reference signal one can also calculate the azimuth of the aircraft. Thus, the precision of the azimuth determination is 0.5°–1° [8].

The DME system operates within the frequency range from 962 to 1023 MHz and assigns frequencies to the specific ground radio relay stations every 1 MHz. The radar signal polarization is vertical. The downlink and uplink frequencies are spaced with 63 MHz. The onboard transmitter and the ground equipment receiver are operating within the frequency range from 1025 to 1150 MHz, and the ground equipment transmitter and the onboard receiver—within the full range. The directional pattern in the azimuth plane is omnidirectional.

The distance is measured by the onboard DME system as follows. The onboard interrogator emits the "distance interrogation" signal—a pair of pulses of a bell form (with the duration at the level of half amplitude of 3.5 μs) with a preset coding

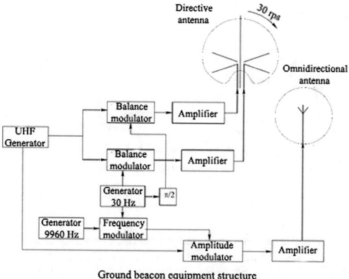

Ground beacon equipment structure

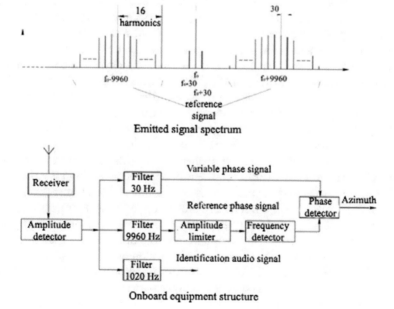

Emitted signal spectrum

Onboard equipment structure

Fig. 10.6   VOR system operating principle

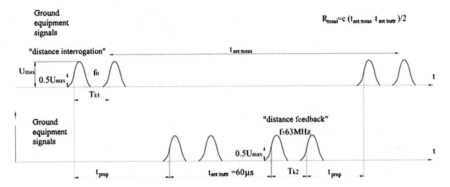

Fig. 10.7 DME system distance measurement principle

distance (12 or 36 µs). The ground radio relay station receives the interrogation signal and (in case the pulse pair decoding is correct) prepares a "distance feedback" signal with a fixed instrument delay. After receiving the feedback signal, within the onboard equipment the time from the moment of half level of the first signal of the interrogation pair is analyzed up to the moment of half level of the first signal of the feedback pair, which finally defines the required distance. Besides the "distance feedback" signals, the ground radio relay station receives the identification signals coded with the Morse alphabet at an audio frequency of 1350 pulse pairs per second, as well as the signals of a chaotic pulse sequence (when there are no "distance interrogation" signals for not less than 700 pulse pairs per second). Initially, the onboard equipment is set to search mode (at that, up to 150 pulse pair per second can be emitted). When it finds the position of the strobes of a receiver (time intervals when "own" signal receiving is expected), which corresponds to the true distance, the system switches to the tracking mode with the interrogation frequency of not more than 30 Hz. The interrogation pulse pairs are wobbled by the emission lifetime in order to avoid simultaneous mutual interference with other aircraft within the coverage area of the radio relay station. The distance measurement principle is shown in Fig. 10.7.

The system provides the onboard distance measurement precision of not less than 200 m. The DME system capacity is limited due to the time principle of signal differentiation. One ground radio relay station can serve up to 100 aircraft. In case this number is exceeded, the most distant from the radio relay station aircraft are excluded from the service (by raising the interrogation signal threshold, to which the feedback must be sent) in order to provide other navigation data users with a required quality of service. Thus, the system flexibility feature is realized [5].

In order to perform the distance measurement during aircraft landing, the precision option of the DME system is used, which allows for a precision of up to 20 m. The DME precision system differs from the above system by pulse form (asymmetrical bell pulse with a steeper edge), processing type (distance measurement performed not by the level of the pulse's half amplitude, but by the initial gradient distance—in order to avoid the multiple reflection influences), and availability of measurement filtering.

10.3.2 National Short-Range Radio Navigation System

The national short-range radio navigation system provides distance and azimuth measurements by one single piece of equipment and within one radio channel. The radio signals of horizontal polarization are formed at frequencies within the range of 726–813 MHz (via downlink) and 837.6–1000.5 MHz (via uplink).

The distance measurement channel operating principle is similar to the operating principle of the DEM system except that the pulses have rectangular form and are coded in triples for downlink interrogations and in pairs for uplink feedback. The distance measurement signals are received and emitted via the omnidirectional antenna. In order to measure the azimuth, a set of signals is formed: "35", "36", and "azimuth signal".

The azimuth signal is a continuous oscillation emitted through a bidirectional pattern (in azimuth plane), which rotates at a rate of 100 rpm. The "35" and "36" signals are coded pairs of pulses, emitted through the omnidirectional antenna at regular time intervals, 35 and 36 times, respectively, per one rotation of a directive antenna. The "35" and "36" signals' position coincides once per rotation of the directive antenna when passing the gap between its directional lobes toward the north and forms the "northern coincidence" signal.

Thus, after the onboard equipment receives the reference "35", "36", and azimuth signals, it is possible to define the azimuth of the aircraft by the delay of the azimuth signal gap relative to the "northern coincidence" signal. The precision of the distance measurement performed by the short-range radio navigation system goes up to 50 m, and the azimuth measurement precision—0.4°.

Besides the main function of the azimuth and distance measurement (on board), the short-range radio navigation system performs aircraft coordinates on board by means of the radio communications "ground indication interrogation" and "ground indication feedback" following the principle of the secondary radar. The azimuth measurement principle within the short-range radio navigation system is shown in Fig. 10.8 [8].

Currently, the most advanced mode of operation of the short-range radio navigation system is being developed. In this mode, the azimuth signal is non-coded pulse emitted every 0.25° of antenna rotation. The distance measurement signals, "35", "36", and "ground indication interrogation," signals are emitted through the directive antenna. This allows simplifying the antenna feeder system of the beacon and the transceiver thanks to reversion from the omnidirectional antennas and emittance of continuous signals.

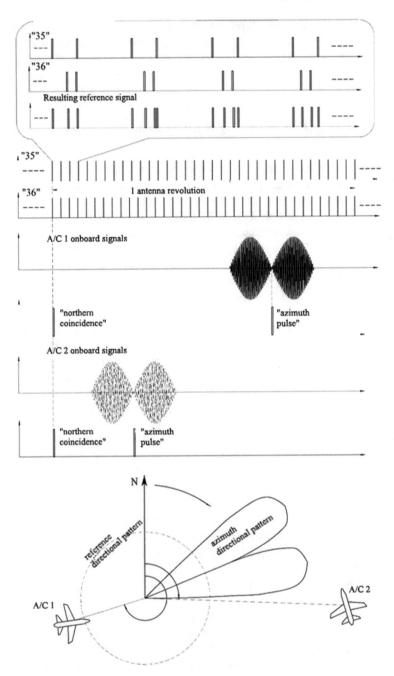

Fig. 10.8 Azimuth measurement by the short-range radio navigation equipment

Besides this, currently activities are performed to transfer the downlink channel from the 700-MHz range to the international navigation frequency range 960–1215 MHz in order to free the frequency band for mobile and digital television systems.

10.4 Global Navigation Satellite Systems

The global navigation satellite systems (GNSSs) are designed for defining the coordinates of fixed objects on the surface of the Earth. GNSS is a combination of ground and space pieces of equipment, which provides radio navigation field and user equipment. The space segment is the artificial satellite system, which moves along the preset trajectory and functions as radio navigation reference points. The satellites move along certain orbits; their number (not less than 24 items) is defined by the need to provide the navigation coverage of the required territory (assuming that each user shall see four satellites as minimum at the moment of coordinate determination). The ground segment consists of the equipment of the command and measurement complex. Its functions are as follows: measurement of the satellite motion parameters, artificial satellite uplink transfer of the required data, as well as management of their equipment operation. The users' equipment consists of the medium frequency (MF) signal receiver and computer, which means that the measurements are conducted in passive (non-interrogative) mode. In order to perform non-interrogative distance measurements, all pieces of equipment (command and measurement complex, artificial satellite system, and users' receivers) contain the universal time system. By the signals of four or more artificial satellite systems, the users' equipment measures the set of the pseudoranges, which differ from the true distances to the satellites because of the difference of the universal time system clock on board of the artificial satellites and within the aircraft onboard equipment, as well as because of the wave propagation error. In order to calculate the geographic coordinates of the aircraft, a system of four equations is solved (10.1):

$$R_i = \sqrt{(x_{\text{art.sat.i}} - x_{\text{A/C}})^2 + (y_{\text{art.sat.i}} - y_{\text{A/C}})^2 + (z_{\text{art.sat.i}} - z_{\text{A/C}})^2} + \delta_P + \delta_{FS},$$
$$i = 1 \ldots 4, \tag{10.1}$$

where R_i are the measured pseudoranges; $x_{\text{art.sat.i}}$, $y_{\text{art.sat.i}}$, $z_{\text{art.sat.i}}$ are the coordinates of the ith artificial satellite; $x_{\text{A/C}}$, $y_{\text{A/C}}$, $z_{\text{A/C}}$ are the coordinates of the aircraft; δ_P is the wave propagation error; δ_{FS} is the error of the users' and space segment equipment frequency standard difference.

The GNSS peculiarity is the moveability of the radio navigation stations. In order to account for this, the artificial satellite system transmits the ephemerides—the data on its orbit parameters corrected by the data received from the command and measurement complex. Thus, after calculating the artificial satellite system coordinates at a given point of time, four unknowns remain in the system of four equations: three aircraft coordinates and the sum of systematic errors. As a result, when four satellites are observed, it guarantees the definition of all aircraft coordinates. The radio navigation signals are the phase-shift-keyed pseudorandom sequences, which allow for acceptable precision at limited signal level [9].

The advantages of the satellite navigation systems are as follows: a vast area coverage (almost the whole Earth surface), high measurement precision, infinite capacity (because of the passive measurement principle), low dependency from the noise interference caused by the propagation anomalies and multiple reflections, terrain relief influence, etc. Several satellite navigation systems exist. The first-generation systems did not have the global and operational efficiency features of the measurements that is why they gave place to the second-generation systems, the most dominant of which are the USA-built Global Positioning System (GPS) and the Russian-built global navigation satellite system (GLONASS).

The GPS system (or NAVSTAR) was developed and is in operation in the USA. Currently, 31 artificial satellite systems are used as intended. The altitude of the circular orbits of the artificial satellite system is 20,200 km, and the orbit revolution period is 11 h and 58 min. The satellites orbit around the Earth in six different planes. The radio signals are emitted at the frequency of 1575.42 MHz. There are two codes: civil and special. In order to avoid the ionospheric error of the navigation measurements, double-frequency measurements are performed, for which the frequency of 1227.6 MHz is used. Since recently, the frequency of 1176.45 MHz is also used. The measurement precision reaches 5 m. In the longer term, the navigation precision can go up 0.6–1.0 m [10].

The GLONASS was developed and initially put into implementation in the Soviet Union; it is now in operation in RF. The space segment includes 24 satellites located in three circular orbits shifted by 120° along the equator (8 artificial satellite systems per each orbit). The orbit altitude is 19,100 km, and the orbit revolution period is 11 h 15 min. The segregation of the signals from different satellites is frequency-type. The system occupies two frequency sub-ranges: 1602–1616 and 1246–1257 MHz [11].

10.5   Instrument and Radio Beacon Landing System

During the approach to landing, the pilot in command (PIC) shall ensure the aircraft movement along the set trajectory with specific tolerances in lateral and vertical directions. Thus, there are two aircraft control tasks: to lateral guidance and glide slope guidance. In order to solve the guidance tasks, it is necessary to have information on aircraft deviations from the set glide slope in two planes. In order to get such information (regardless of the weather conditions), the radiotechnical aids are widely used. There are two approaches to the navigation service provision during landing. In the first case (the radio beacon landing system used), the definition of the aircraft coordinates relative to the runway and the calculation of the deviations from the heading and the glide slope are performed on ground, and the control commands to the flight crew are passed through the radio channel (usually, these are voice commands). In the second case (use of the radio beacon landing system and the instrument landing system), the current aircraft position is defined on board relative to the required trajectory based on the processing of the signals from the radio navigation field created by the ground radio beacons. The use of two systems with different operating principles increases the reliability of landing operations [12].

10.5.1 Approach Radar System

The approach radar system consists of the dispatch surveillance radar equipment (SRE) and precision approach radar (PAR), as well the direction finder and radio communications aids. The dispatch SRE is functionally similar to the ATC SREs and serves for aircraft guidance up to the border of the PAR coverage area. The direction finder serves as a redundancy or standby aid. The main element of the approach radar system is PAR, which provides vertical and lateral aircraft deviations from the set glide slope and distance from the aircraft to the runway threshold (it can be used in a stand-alone mode—without other radio beacon landing system aids). The measured deviations are read by the operator and passed through the uplink channel via the radio communication aids. The advantage of such system is that in this manner the landing information service provision can be arranged for any aircraft regardless of its equipment (only the radio communication equipment is required). The landing direction finders are, as a rule, operating in the three-centimeter wave band (operation frequency of 9000–1000 MHz), which allows creating antennas with narrow directional patterns and getting good resolution and precision of measurements. The main channel of PAR is primary, which means that it operates by signals reflected from the aircraft. In order to measure the directional coordinates, two channels are used: localized and glide slope channels. The measurements are conducted by means of scanning of the beam directional patterns. The directional pattern of the localizer antenna is narrow in the horizontal plane (about $0.8°$) and wide in the vertical plane (about $6°$). The directional pattern of the glide slope antenna is wide in the horizontal plane (about $20°$) and narrow in the vertical plane (about $0.6°$). The PAR is conducting the beam swinging (digital in case of phased arrays, and mechanical if the reflector-type antenna) with the frequency of 0.5–1 Hz in sectors up to $35°$ by azimuth and up to $9°$ by elevation. The range of a standard landing radar is 20–30 km. The landing radar and service sector orientation are shown in Fig. 10.9 [13].

The errors of the directional coordinates' definition (re-calculated into the linear deviations from the glide slope) do not exceed the sum of the tenth of the linear deviation from the glide slope by the measured parameter and half percent of the distance. The error of the distance measurement does not exceed 30 m, which is ensured by the use of short main pulses (about 0.5 μs). Within the landing radar either one set of transceiver and processing equipment (then it is commutated between the localizer and glide slope antennas), or two sets—each for its own channel. The standard position of the landing radar is close to the runway, at a distance of 120–185 m from it, and further to the runway center—7000 to 1200 m from the aircraft touchdown

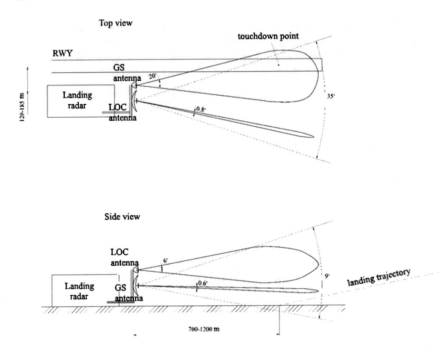

Fig. 10.9 Location and coverage area of the landing radar

point. Depending on the specific installation point, the angular position of the radar is chosen—in such a manner that one touchdown point is within the field of coverage and approximately equivalent coverage areas to the right and to the left from the heading plane are ensured. In order to increase the coverage area, the secondary radar principle is used, which means that the aircraft equipped with the transponders are forming the active feedback signal.

In order to counteract the passive interference, MTI and radio signal polarization inversion are used. At the workstation of the landing operator, the heading and glide slope indicators are available, at which the images in "angle–distance" or "deviation–distance" coordinates are formed. The display scan by the directional coordinates is linear, and by the distance—logarithmic. The indicator view is shown in Fig. 10.10.

10.5.2   Reduced Landing Systems

The reduced landing system is designed as a combination of markers located at the extension of the runway centerline with certain spacing intervals.

In our country, 1 km from the runway threshold the middle locator and locator middle marker (LMM) are set, and 4 km from the runway threshold the outer locator

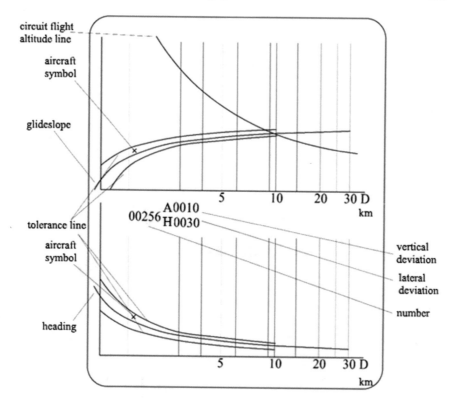

Fig. 10.10 Landing radar indicator

and locator outer marker (LOM) are set. In accordance with the international standards, the LMMs are set at a distance of 1 km from the runway threshold, and the LOMs are a distance of 7 km (besides this, in some cases the locator inner marker is set at a distance of 400 m from the runway threshold).

The locator beacons are emitting continuous oscillations at the frequencies of 190–1750 kHz. On board, these signals are received by the antenna of the automatic direction finder (ADF), which defines the direction of the incoming radio wave front, i.e., the relative bearing (RB) of radio station. The RB value is indicated on the dial-and-pointer indicator. By comparing the directional angles of the middle and outer locators, the pilot can define the deviation from the runway heading. In order to identify the locator beacons, the manipulation emitted with the Morse code oscillations with the audio frequencies of 1020 or 400 Hz. These guidance stages are shown in Fig. 10.11.

In order to provide the aircraft with the glide slope guidance, the markers are used. The markers emit a continuous carrier with a frequency of 75 Hz, which is modulated with the audio frequencies of 3000, 1300, or 400 Hz. The audio frequency's manipulation code and its value depend on the location of the marker. The direc-

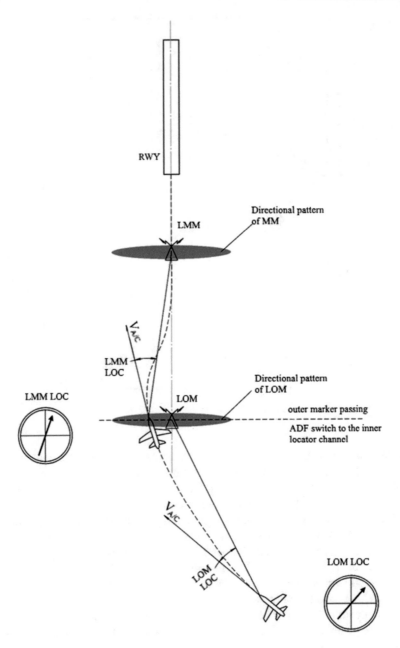

Fig. 10.11 Maneuvering in horizontal plane at landing

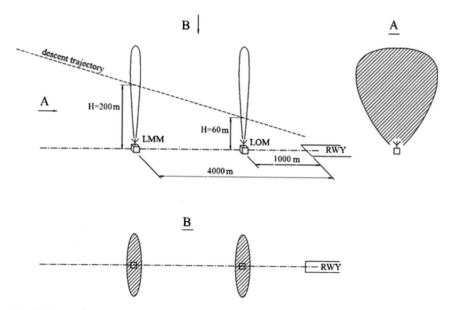

Fig. 10.12 Position of the markers

tional pattern of the marker is directed upward. It is narrow in the vertical plane (that passes along the runway) and wide in the vertical plane (that is perpendicular to the runway). Thus, when flying along the heading and passing above the locator marker, the marker receiver on board receives its signal, and the pilot checks the distance to the runway threshold at this point of time (the marker's position is known). The use of the reduced approach system is shown in Figs. 10.11 and 10.12 [14].

10.5.3 Radio Beacon Landing Systems

The radio beacon landing systems consist of two radio beacons, whose emission creates a signal field; when receiving these signals, it is possible to define deviation from the heading plane and glide slope plane of the landing trajectory. There are radio beacon landing systems of meter, decimeter, and centimeter wave bands. The most widely used systems are the meter (instrument landing system (ILS), regulated by the international standards) and the decimeter (radio beacon landing group, which is used in our aviation) bands.

The LOC is positioned at the extension of the runway centerline past the outer runway threshold; the glide slope beacon is positioned to the side of the runway at the level of the inner runway threshold. The distance within the ILS is defined with the additionally installed radio relay station of the DME (positioned at the location of the localizer) or markers located in a manner described in item 10.5.2.

The distance within the radio beacon landing group system is defined by the radio relay station of the distance measuring equipment, which is operating at the frequency and following the principles of the distance measurement channel of the short-range radio navigation system (positioned jointly with the localizer). The radio beacon landing systems are assigned with categories.

The Cat. I systems allow for automatic approach to landing up to the height of 60 m with the runway visibility of not less than 800 m; the Cat. II systems—up to the height of 30 m and runway visibility of not less than 400 m; and the Cat. III systems—up to the height of 15 m and runway visibility of not less than 200 m (Cat. IIIA), up to the touchdown and runway visibility of not less than 50 m (Cat. IIIB), and up to the full stop, including the taxiing, with zero runway visibility (Cat. IIIC).

The ILS localizer creates a horizontally polarized field at the frequencies of 108–111.975 MHz (with the carrier frequency separation of 5–14 kHz) amplitude-modulated with the frequencies of 90 and 150 Hz. During approach to landing, from the right-hand side from the heading plane the high-frequency carrier modulation with the tonal signal of 150 Hz prevails, and from the left-hand side—of 90 Hz. Thus, within the heading plane the equisignal direction is created, and with the help of the difference in factors of modulation of the received signal with the frequencies of 90 and 150 Hz, the track angle error (from the heading plane) can be defined. Besides the main function—aircraft heading guidance—the localizer performs the identification function, which is done by a manipulation of the carrier by the Morse code with the note frequency of 1020 Hz. The localizer coverage is 45 km within the ±10° sector from the heading, 30 km within the ±35° sector, and 18 km outside the above-mentioned sectors. The localizer operation is shown in Fig. 10.13.

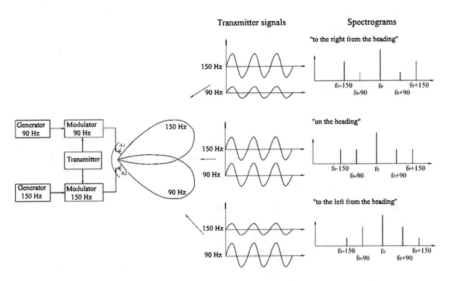

Fig. 10.13 Operating principle of the localizer

The ILS glide slope marker creates a horizontally polarized emission at the frequencies of 328.6–335.4 MHz (with the spacing of 4–32 kHz) amplitude-modulated with the frequencies of 90 and 150 Hz. Above the glide slope plane, the modulation with the frequency of 90 Hz prevails, and below the glide slope plane—with the frequency of 150 Hz. Similar to the heading guidance, after the signal processing by the onboard glide slope receiver, the track angle error (from the glide slope plane) is defined. The glide slope marker coverage area is equal to $\pm 8°$ within the azimuth sector, from 0.45θ to 1.75θ within the elevation sector (θ is the glide slope angle; usually, it is equal to $3°$), and about 18 km in distance. The most critical aspect is the formation of the directional pattern of the glide slope marker in the vertical plane.

The difficulty is that the form of the pattern significantly depends on the ambient terrain relief: The direct signal (toward the aircraft) and the signal reflected from the ground interfere, the resulting directional pattern becomes multi-lobed, and the higher the antenna is elevated, the more unevenly the directional pattern is shaped. As it is very difficult to form a directional pattern with a steep lower front in order to avoid the reflections from the ground, the antenna system of the glide slope beacon is calculated subject to the influence of the ground surface. Antennas are located in such a way to form the equisignal direction at a required glide slope angle which is formed due to the interaction of the first interference lobes of the upper and lower antennas. The negative aspect of such choice is that a thorough care for the surface located before the beacon antenna system is required, as the change of its condition (plant formation, snow, etc.) will lead to the glide slope change [15]. The glide slope beacon operating principle is shown in Fig. 10.14.

The higher is the landing system category, the higher are the requirements to the radio navigation field parameter provision precision, as well as to the reliability and uninterrupted provision of data to the aircraft.

Besides this, within the Cat. III systems additional measures shall be implemented: The glide slope leveling beacon (aircraft movement support below the height of 15 m (when the aircraft is above the runway threshold) and up to the touchdown), the glide slope beacon and localizer of return direction (in order to ensure navigation service of the go-around), as well as the localizer and the glide slope beacon shall have a clearance channel. The clearance channel is designed to create a radio navigation field in order to inform the pilot on the direction of aircraft deviation from the sector of proportional guidance (of the localizer or the glide slope beacon).

The radio beacon landing group system also consists of the localizer and the glide slope beacon positioned similar to the ILS. The radio beacons also create the horizontally polarized emission with the amplitude modulation of different frequencies to the different sides of the glide slope and heading planes. The localizer emits signals within the frequency range of 939.6–966.9 MHz, and the glide slope beacon emits radio signals within the range of 905.1–932.4 MHz. The difference from the ILS is that the equisignal direction is formed not by the constant emission of the signals with different modulation parameters, but by means of switching the antennas with

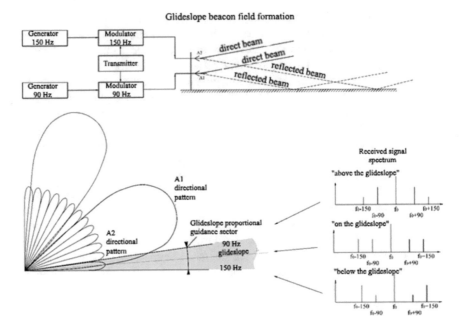

Fig. 10.14 Operating principle of the glide slope beacon

equal time intervals (the right and the left one for the localizer, and the upper and the lower one for the glide slope beacon); besides this, the modulation is performed not as a continuous signal, but by rectangular pulses with the frequency of 2100 Hz for one antenna and 1300 Hz for the other antenna [16].

10.5.4   Landing Procedure Using the Satellite Navigation Systems

The characteristics of the standard satellite navigation system modes (see item 10.4) cannot ensure the precision required for the aircraft landing. In order to increase the precision of the coordinates' determination with the help of the satellite navigation systems, the differential mode is used; this mode is implemented by adding corrections to the data on the navigation satellites' ephemerides. The corrections are formed by a specialized ground equipment (monitoring and correcting station) and passed on board through a one-way VHF radio line. The differential mode principle is as follows: On ground, at the point with the coordinates known with high precision, a receiving computer of the satellite navigation system is located. In the simplest case, the coordinates measured with this receiving computer are compared to the known (true) coordinates and the measured difference is passed through the radio channel to all users concerned. But it is finally possible that within the user's receiving computer a different working constellation (combination of navigation satellites, whose signals are used for the measurement) will be chosen. At that, the systematic positioning error can differ from the value measured by the monitoring and correcting station, and the compensation procedure becomes non-precise. In order to avoid such situation, another method for the differential mode implementation is used: when the monitoring and correcting station passes the corrections to the measurement of the pseudoranges of each satellite. Such correction method is invariant to the satellite choice.

In order to perform the precision approach to landing with the help of the onboard satellite system, it is necessary to have a correction receiving channel and navigation data processing algorithms (subject to these corrections) [17].

Concluding the study of the radiotechnical air traffic support aids, we will list the sequence of their use subject to the stages of the flight.

During taxiing and rolling, the controller uses the airport surveillance radar (if available), and the PIC uses the visual references.

During rolling and takeoff, the pilot uses the self-sustained navigation aids (radio altimeter and pressure altimeter), visual references, and (when the radio beacon landing systems are designed with the return direction) readings of the respective instruments. At this stage, the controller uses the airport surveillance radar (primary, secondary, or the integrated system complex, which includes the primary and secondary channels) or surveillance radar of the radar landing system. During the enroute stage (at the preset level), for the positioning the flight crew uses either the short-range navigation systems or satellite systems or (as a standby navigation aid) the orientation by the beacons with the help of the direction finder. Where in case of utilization of the VOR system jointly with the DME or the short-range radio navigation system, the flight along an arbitrary path is possible, and in case of utilization of the VOR system only or one beacon, the flight is performed to the beacon or from the beacon, i.e., between the reference navigation points. At this flight stage, the controllers use the enroute surveillance radars. The altitude is defined on board by the self-sustained aids; the controllers receive it in the feedbacks from the secondary radar.

During flight outside of the radar or short-range navigation system coverage, the flight crew determines the aircraft position based on the wide-ranging or satellite navigation aids. When there is no possibility to interact with the outer navigation systems, the flight is performed based on the data received from the self-sustained systems (the speed—from the air data computer system; the altitude—form the pressure altimeter and radio altimeter; the drift angle—from the radar indicator; and the geographic coordinates—from the inertial system). When signals from a more precise navigation system appear, correction of the self-sustained data is performed or switching to this navigation system is preformed provided the radio contact is stable. Thus, the inertial system or the wide-ranging navigation system can guide the aircraft to the short-range navigation system or the ATC radar coverage area. Using the short-range navigation systems or following the controller's commands (based on the data from the surveillance radars), the flight crew can guide the aircraft to the airport surveillance radars and landing system coverage area.

Depending on the airport equipment, the landing is performed based on the microwave landing system (MLS), ILS, or radio beacon landing group (when on board, the localizer and the glide slope beacon receivers are used, as well as the aircraft distance measurement equipment or marker receiver). When there are no marker aids, the landing is performed with the use of the markers following a reduced pattern (the use of the marker receiver on board, automatic direction finder, and radio altimeter). In case of the on board equipment failure, the approach to landing with the instrument and reduced landing systems is performed with the help of the guidance commands received from the radar landing system operator (on board, only the voice radio communication system shall be serviceable).

During the biggest part of the flight, the systems, which define the coordinates of the aircraft on ground (precision approach radar, airport surveillance radar, enroute surveillance radar, automatic direction finder), complement and to a certain degree duplicate or provide redundancy to the aids that help performing positioning on board (short-range radio navigation system, self-sustained aids, landing systems) increasing the reliability and flight safety.

LOM—locator outer marker, OM—outer marker, LMM—locator middle marker, MM—middle marker, IM—inner marker, GS beacon—glide slope beacon, LOC—localizer, azimuth–DME—azimuth–DME ground beacon of the short-range radio navigation system, RWY—runway, START—start air traffic control unit, VHF radio—very-high-frequency radio, RADAR—radar landing system, ADF—automatic direction finder, METEO—weather radar, ASR—airport surveillance radar, A/D SR—aerodrome surveillance radar, TOWER—ATC tower, enroute SR—enroute surveillance radar.

A particular place is given to the satellite navigation systems. In case of a reliable radio contact with the required number of navigation satellites, the precision of the coordinate measurement, as a rule, exceeds the precision provided by other systems, which is true for almost all stages of the flight. Thus, the use of the satellite systems usually has the highest priority. The controller himself/herself uses the satellite navigation system data rather indirectly: When on ground and on board, the ADS system is available, which functionally and to a large extent dominates other ground surveillance systems [6].

The standard positioning of the radiotechnical flight support aids is shown in Fig. 10.15.

Section Review

1. Tactical and technical characteristics of aerodrome and route surveillance radars.
2. Differences of national and international modes of secondary radars.
3. Key points of the automatic dependent surveillance concept.
4. Key features of the VOR/DME system.
5. The structure of the signals of national short-range radar navigation system.

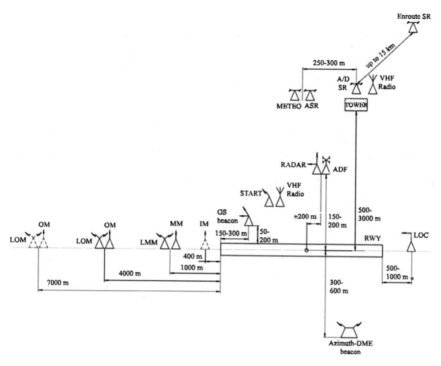

Fig. 10.15 Positioning of the radiotechnical aids within the A/D area

6. Basics of the global satellite navigation systems.
7. Precision approach radar operating principle.
8. Approach to landing principles with the help of the locator markers.
9. Basics of the aircraft marker landing systems.
10. Location of the ground radar, navigation, and landing system aids. Match the use of the specific radiotechnical systems with the aircraft movement stages.

References

1. Akhmedov RM (2004) Automated ATC systems. Polytechnica, St. Petersburg, 440 p
2. Tuchkov NT (1994) ATC automated systems and radio aids. Transport, Moscow, 363 p
3. Kuznetsov AA, Dubrovskyi VI, Ulanov AS (1983) ATC means operation. Reference book. Transport, Moscow, 256 p
4. Agadzhanov PA, Vorobyov VG, Kuznetsov AA, Markovich ED (1980) Automation of airplane navigation and air traffic control. Transport, Moscow, 357 p
5. Yarlykov MS, Boldin VA, Bogachev AS (1980) Radio navigation units and systems. ZhAFEA, 383 p
6. Khaymovich IA et al (1990) Under the editorship of Sosnovskyi AA. Radio navigation. Reference book. Transport, Moscow, 264 p

7. Manual on the Secondary Surveillance Radar (SSR) Systems. Doc 9684 AN/951 ICAO
8. Bondarchuk IE (1978) Flight operation of the navigation equipment. Transport, Moscow, 272 p
9. Shebshaevich VS (1993) Radio navigation network systems. Radio e svyaz, Moscow, 387 p
10. Annex 10 to the Convention on International Civil Aviation. Aeronautical telecommunications. Volume I—Radio navigation aids
11. Kharisov VN, Perov AI, Boldin VA (1988) GLONASS global navigation satellite system. Reference book. EPRZh, Moscow, 387 p
12. Belgorodskyi SL (1972) Automation of landing systems. Transport, Moscow, 352 p
13. Larionov LA, Martynov YG, Spiglazov YA, Zhuravlev AI (1977) Radar landing systems. Voyenizdat, Moscow, 240 p
14. Makarov KV, Tchervetsov VV, Sheshin IF, Volynets VA (1978) Airport Radio navigation aids. Transport, Moscow, 336 p
15. Shatrakov YG (1982) Radar landing systems using an angle measuring equipment. Transport, Moscow, 160 p
16. Babay GA, Pavskyi AG (1956) Radio navigation aids. Moscow, 180 c
17. Korol VM, Shatrakov YG (2011) Basic radio navigation. Learning guide. StUAI, St. Petersburg, 106 p

Chapter 11
Economics at ATC Automation Systems' Implementation

11.1 Economics at Passenger and Cargo Transportation

Within the civil aviation transportation system, the end-users of the air transport (passengers or cargo forwarders and consignees) purchase the specific services from the airlines—organizations that operate the aircraft. But in the frame of the aviation transportation, infrastructure significant expenses are also attributed to the airports and air navigation services, which are usually provided by other organizations. In order to compensate expenses due to the air traffic management and services, charges are applied. Thus, the financial chain is looped: Air traffic services' users pay for the flight itself, and the airlines pay the charges to the ground service providers. And depending on the form of ownership of the air traffic service companies, the payments can be both direct and indirect (e.g., through the state) [1].

ICAO has developed a specific policy on interaction between the airlines, airport and air navigation services, and public sector subject to setting and payment of charges. Provisions of this policy are stipulated in a number of officially published documents [2–4]. Key aspects regulated by ICAO in the field of financial cooperation between the organizations participating in the air services provision are aimed to ensure:

- priority of the flight safety requirements—principle of independency of the flight safety from financial aspects;
- possibility to control expenses of the organizations providing airport and air navigation services—principle of transparency of setting the charges;
- commitment to prevent monopolization of air traffic services provision and thus, raising the charges—principle of reasonable profit formation (as a rule, connected to the absence of the free competition);
- avoidance of price discrimination subject to charging the airlines by a certain criterion—principle of equality.

In order to implement the above-mentioned principles, an adequate state monitoring is encouraged.

Bestugin A.R. et al., *Air Traffic Control Automated Systems*,
Springer Aerospace Technology, https://doi.org/10.1007/978-981-13-9386-0_11

The cost basis for the charges is set either proportionally to the number of landings (e.g., approach control charges), or proportionally to the weight of the aircraft (landing charges), or proportionally to the distance flown within the air navigation service area (air navigation charges), or based on a more complicated dependency. In any case, during setting charges the principle of justice is applied—amount of charges of each organization is dependent on the volume of transportation or on the expenses, to which the charges lead, from the side of the operator of the ground services.

Out of the numerous charges that the airline shall pay, the following can be outlined as they are directly connected to air traffic management and services:

- landing charges (including the lighting aids and control charges);
- aerodrome control charges;
- approach control charges;
- route air navigation service charges;
- flight planning and flight information service (FIS) charges;
- communication aids charges (ground and/or satellite-based facilities supporting aeronautical fixed service (AFS), aeronautical mobile service (AMS), and aeronautical broadcast service);
- navigation charges (ground-based radio and visual aids to navigation, global navigation satellite system (GNSS) and its associated augmentation systems in support of all phases of flight);
- surveillance charges (primary/secondary surveillance radars and other ground/satellite-based facilities supporting the automatic dependent surveillance (ADS) and/or ADS-broadcast (ADS-B));
- meteorological charges.

It shall be noted that in every situation, different options for the cooperation between the stakeholders are possible. The charges can be paid separately by each item or the charges can be combined so that some charges are included into other charges. Besides this, air navigation service providers and airports can use reciprocal payments subject to charges (if it is mutually convenient). In other words, specific details can be defined with a certain level of freedom.

In Fig. 11.1, a schematic procedure of setting air navigation charges and airlines charging is provided.

It is remarkable that when the scopes of work of the airlines, airports, and air navigation service providers are separated, no tendency to conservativism is observed (unwillingness to implement new systems or expand the service provision due to the associated significant expenses). The reason is that the coverage area and the quality of the air traffic service provision define the main criterion of the "product" of the ground services—traffic capacity. Thus, the expenses associated with the commissioning of advanced technologies and expansion of the scope of work of the flight service systems are compensated with the possibility to raise the charges paid by the air carriers that will get certain benefits, for example:

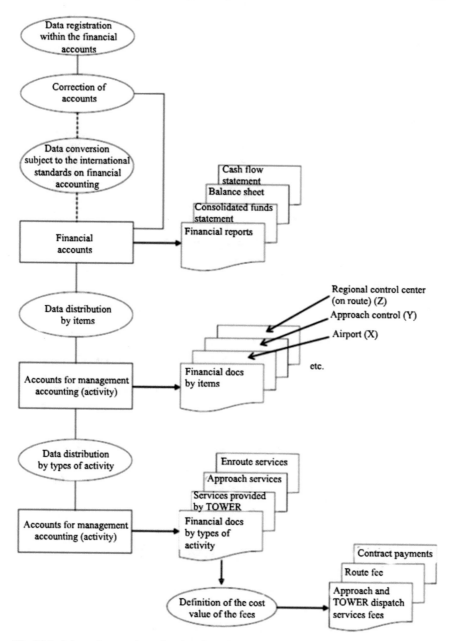

Fig. 11.1 Schematic procedure of setting air navigation charges and charging

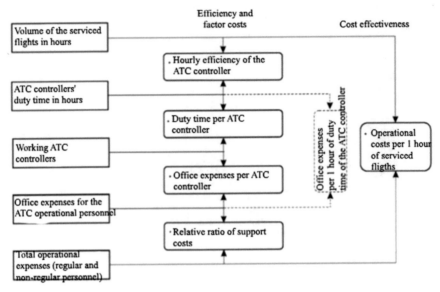

Fig. 11.2 Use of cost performance and target products' figures for cost-effectiveness assessment

- possibility to choose the routes for the flights;
- reduction of the number of delays in flight due to the optimal air traffic management;
- reduction of the number of delays due to consideration of the meteorological conditions;

These benefits lead to direct financial results, and thus, the financial chain is looped again. In Fig. 11.2, a structure for assessing the cost effectiveness with the help of the cost performance indicators is provided.

- possibility to perform flights in more complicated meteorological conditions;
- increase of the flight safety level.

11.2   Cost Effectiveness Associated with the Use of the Simulators

Implementation of simulators (Simulation Training Devices (STDs)) into the ATC controllers' training and re-training process plays increasingly larger role as they turned out to be a cost-effective means for providing the controllers with the most vital element of the training process—the experience. The main option to provide a high-quality training of an ATC controller is to let him/her gain practical experience. Such experience can be gained in three ways: during air traffic control provision in real operation environment under supervision of an instructor; during use of real aircraft control equipment (provided for practical training of the ATC controllers only); with the help of STDs that allow working out the skills without real air traffic control provision. The first way can be associated with different types of the flight safety risk and other issues due to the difficulties to work out the tasks in abnormal operations. The second way is extremely expensive and far from the most effective. From the financial and safety points of view, the use of ATC controllers' STDs is the most effective option for the ATC controllers' practical training.

Analysis of the experience gained at ATC controllers' STDs development and implementation allows distinguishing several groups of parameters that are appropriate to be used for the STDs' cost-effectiveness assessment.

A. The group of the parameters that characterize the controllers' level of training.

1. performance of certain operations (detection, distinction, etc.), adoption of decisions (tracing, framing of commands, etc.), performance of functional tasks (calculation, physical motor operations, use of the keyboard or of the manipulator, etc.);
2. performance of integral operations, i.e., of certain elements of exercises and tasks that require the adoption of decision on the sequence of actions (including stated above), and different simple operations;
3. performance of complex operations (functions) that require ability to combine chosen sequences of simple operations;
4. performance of complex functions in abnormal situations (prohibitions, restrictions, abnormal conditions, etc.);
5. performance of complex functions in in-flight emergencies (when the controller is subject to stress).

B. The group of the parameters that characterize the STDs' functional abilities.

1. number of students to train simultaneously;
2. time (duration of the play) of the exercise;
3. number of simultaneous exercises;
4. time of use (available time) of the STD within a certain period of time;
5. total number of the STD-trained specialists within a certain period of time;
6. correspondence of the simulated tasks with the real operating environment of the controller;
7. number of simultaneously simulated aircraft, and their types;
8. ATC areas simulated with the STD;
9. types of displayed information and its correspondence with the reality;
10. possibility of joint and separate operation of the STD modules.

C. Economic parameters.

1. price of the STD;
2. area required for its allocation;
3. number of maintenance staff;
4. use of additional equipment during adjustment and alignment at the stage of software development and testing;
5. maintenance and operation expenses during controller's training.

D. The group of the parameters that characterize the controllers' training efficiency.

1. availability of the computer-aided monitoring of the training process;
2. availability of the adaptive control of the training process;
3. the rate of growth of the controller's level of training within a certain period of training time;
4. duration of the on-the-job training prior to the unsupervised work after the STD training.

The planned efficiency is the main factor for adoption of decision on the purchase and integration of the ATC controller's STDs. When using a technical aid, beside the parameters that characterize the controllers' level of training (group A) and the STDs' functional abilities (group B), the associated financial expenses (group C) must be considered in order to minimize such expenses. The below method of efficiency comparison for different options of STDs implementation directly and indirectly takes into account the parameters of the groups A, B, C, and D, except for the item 7 of the group B. In order to conduct comparative assessment of the expenses, it is reasonable to introduce the notion of the price of the training when using the technical aid. It is obvious that the price of the training P_{tr} will be higher if the operation costs of the technical aid are higher (taking into account its life):

$$P_{tr} = P_{y\,hour} \cdot T_{tr}, \tag{11.1}$$

where $P_{y\,hour}$ is the price of use of the specific STD per one hour;

T_{tr} is the time needed for solving of a certain round of tasks (exercising of certain practical skills and abilities for which the specific STD is used).

The time of the training can be considered as a sum of the t_{Ni} (defined by a certain training program) needed for exercising the tasks given in items 1, 2, 3, and 4 of the group A:

$$T_{tr} = \sum t_{Ni}, \tag{11.2}$$

where t_{Ni} is the time needed for the development of a skill or an ability and marked with a sequential number.

The price of the use of an STD of a certain time per one hour is comprised of operational expenses P_{op} and capital costs P_{cap} associated with the facilities provision purchase of the equipment taken relative to the time of the potential use till the end of life T_1:

$$P_{y\,hour} = P_{op} + P_{cap}/T_1. \tag{11.3}$$

The operational expenses reduced to the unit of time include:

– price of electrical power;
– labor expenses for the operational and maintenance staff;
– payments for the facilities taking into account the depreciation costs;

– price for the scheduled maintenance and repair.

The above method of training expenses assessment can be realized if the STD is constantly in use during the whole period of training without the down time.

However, in practice the STD is used at intervals, and the full training price will be defined as:

$$P_{tr\,full} = P_{tr} + P_{down},\tag{11.4}$$

where P_{down} is the price for the STD down time.

If the price for the down time per hour is $P_{y\,down/hour}$, then the full down time price will be:

$$P_{down} = P_{y\,down/hour} \cdot T_{down}.\tag{11.5}$$

The down time can be expressed as the difference between the time of the potential use of the STD T_{pot} and the time of the real use of the STD T_w defined by the instructional program of the training process:

$$T_{down} = T_{pot} - T_w.\tag{11.6}$$

In order to assess the amplitude of the efficiency of the STD use, a usage coefficient can be introduced:

$$K = T_w / T_{pot}.\tag{11.7}$$

Then, the down time of the STD can be expressed as:

$$P_{down} = P_{y\,down/hour}(1 - K)T_w/K.\tag{11.8}$$

The full price of the STD training will be:

$$P_{tr} = P_{y\,hour}T_w + P_{y down/hour}\frac{1 - K}{K}T_w.\tag{11.9}$$

In order to compare the price of the training process when using the STDs of different capacity (different number of workstations for the students to train simultaneously), it is necessary to perform calculations in accordance with the present methodology reduced to one student. Thus, it is reasonable to introduce the notion of the price of the training and the price of the STD down time for one hour per one student:

$$P_{y\,use} = \frac{P_{y\,hour}}{n} \text{ and } P_{y\,down} = \frac{P_{y\,down/hour}}{n},\tag{11.10}$$

where $P_{y\,use}$ is the price of the full STD use for one hour reduced to one student;

$P_{y\,\text{down}}$ is the price of the STD down time for one hour reduced to one student;
n is the number of students that are training with this STD simultaneously.

In this case, the price of the training for one student on a specific STD (taking into account the possible down time) will be defined as:

$$P_{\text{tr}} = P_{y\text{use}} \sum t_{N_i} + P_{y\text{down}} \frac{1-K}{K} \sum t_{N_j}. \tag{11.11}$$

If in accordance with the required training program several STDs must be used, the overall training expenses can be considered as a sum of expenses calculated with the Formula (11.11) for each STD separately and taking into account the amount of time that will be spent at each STD; finally, we will get the following formula:

$$P = \sum_{y=1}^{m} P_{y\text{use}} \sum_{j=1}^{n_y} t_{N_{yj}} + \sum_{y=1}^{m} P_{y\text{down}} \left(\frac{1-K_y}{K_y} \right) \sum_{j=1}^{n_y} t_{N_{yj}}, \tag{11.12}$$

where C is the price for the full training by N skills (abilities) with the use of m types of STD modules reduced to one student;

$P_{y\text{use}}$ is the price of the use of one type of an STD module for one hour reduced to one student;

$P_{y\text{down}}$ is the price of one type of an STD module down time for one hour reduced to one student;

$t_{N_{yj}}$ is the time required to develop a skill or an ability N_j when training on an STD module y;

$$K_y = \frac{T_{yw}}{T_{y\text{pot}}}$$

where T_{yw} is the time required to develop skills and abilities trained at an STD module type y;

$T_{y\text{pot}}$ is the time of maximum possible use of an STD module type y;

m is the number of STD module types;

n_y is the number of the skills and abilities that are developed with an STD module type y

Thus, the parameters that characterize the STDs' functional abilities subject to the financial efficiency (group B) except for items 5 and 7 are reflected in formula (11.12) as follows.

Item 1 in the sense of n;
      2 in the sense of $t_{N_{yj}}$;
      3 in the sense of m;
      4 in the sense of T_w;
      6 in the sense of m;
      8 in the sense of m;

9 in the sense of m;

10 in the sense of m.

Economic parameters of the group C are reflected as follows.

Item   1 in the sense of P_{cap};
        2 in the sense of P_{cap};
        3 in the sense of P_{op};
        4 in the sense of T_w;
        5 in the sense of P_{op}.

The parameters that characterize the controllers' training efficiency (group D) are reflected as follows.

Item   1 in the sense of $t_{N_{yj}}$;
        2 in the sense of $t_{N_{yj}}$;
        3 in the sense of T_{pot};
        4 in the sense of T_w.

This methodology allows for assessing the price of different training processes based on the use of different types of STDs with the STD training times as per the instructional programs. The higher the precision of the input parameters (price of operation and down time for each STD type; time of use and down time within the training process; time required to develop skills and abilities in accordance with the instructional program) is, the more precise the performed assessment will be. In this respect, it is reasonable to introduce and use integral STDs built by modules that allow for configuring a number of universal STD modules. At this, the universal STD modules can simulate different ATC sectors and play independent exercises of a preset complexity level. The number of the universal STD modules within one integral STD is defined by the capacity of the controller training center and its instructional programs. If required in accordance with the instructional programs the universal STD modules can be configured into one single system and provide for simulation of the ATC within a preset interconnected combination of sectors.

Finally, these are economic parameters that characterize the overall STD implementation efficiency (upon condition of providing for the development of all required skills and abilities by the ATC controllers during their practical training at STDs) [1].

References

1. Velkovich MA, Shatrakov AY et al (2011) Financial safety and stability of enterprises, Economics, Moscow, p 376
2. ICAO's policies on charges for airports and air navigation services. Doc 9082 ICAO
3. Manual on air navigation services economics. Doc 9161 ICAO
4. Airport economics manual. Doc 9562 ICAO

Printed in the United States
By Bookmasters